ABANDONED HUDSON VALLEY

EXPLORING NEW YORK'S FORSAKEN PAST

LIZ COOKE AND ANDY MILFORD

AMERICA
THROUGH
TIME

America Through Time
www.through-time.com

First published 2025
Copyright © Liz Cooke and Andy Milford 2025

ISBN 978-1-63499-508-5

Typeset in 10pt on13pt Trade Gothic
Printed and bound in England

CONTENTS

INTRODUCTION

In the fall of 1825, painter Thomas Cole took a steamship ride up the Hudson River in New York State. He did not know what he would find since so much of the upper Hudson was still wild. He stopped first at West Point, the rocky bluff overlooking the Hudson River that is home to the United States Military Academy. He then went further north until he reached the Catskill Mountains. It was there that he found his inspiration. His work, and that of his contemporaries, became known as the Hudson River School of Painting. He had discovered what indigenous people, explorers, and locals already knew. This place is magic.

When people today think of the Hudson Valley, images come to mind of country roads and county fairs, long stretches of river and high mountain peaks, rustic landscapes steeped in the colors of fall, and a blanket of white snow in the depths of winter. Natural beauty and a vibrant way of life are why so many people call the Hudson Valley home. Quite honestly, it is what drew us here as well. With its quirky, bohemian villages and diverse, modern cities, the Hudson Valley is a place for new beginnings and old traditions. It is a lovely place with an extraordinary history. In fact, much of America's history was written in its towns and villages. For some, the area's recorded history is fascinating enough, but for others—like us—the untold stories are far more compelling.

On these pages, you will see a different side of the Hudson Valley. You will see places like the once-hopping resorts of the Catskills (known as the "Borscht Belt") that are deteriorating in plain sight. You will see inside the ruins of once-active psychiatric institutions, harkening back to a time before deinstitutionalization, when each of these places was a city unto itself. You will see inside abandoned homes and mansions with furnishings perfectly intact and personal effects left neatly behind.

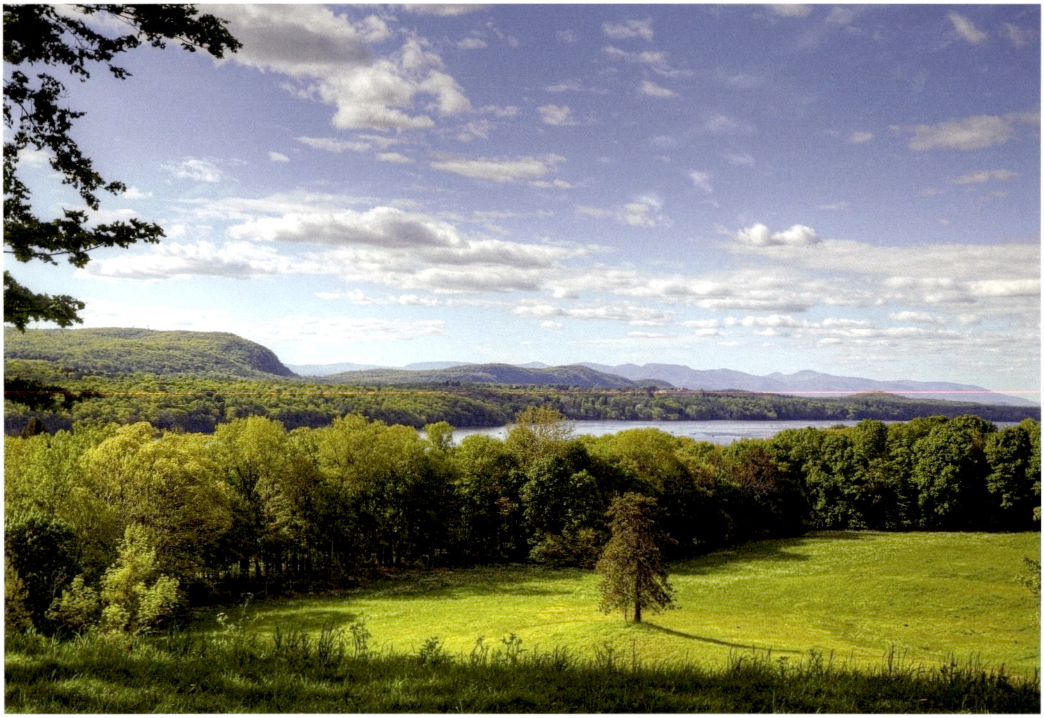

Generations have been enchanted by the natural beauty of New York's Hudson Valley. Seen here, the Hudson River and Catskill Mountains. (*Photo: Andy Milford*)

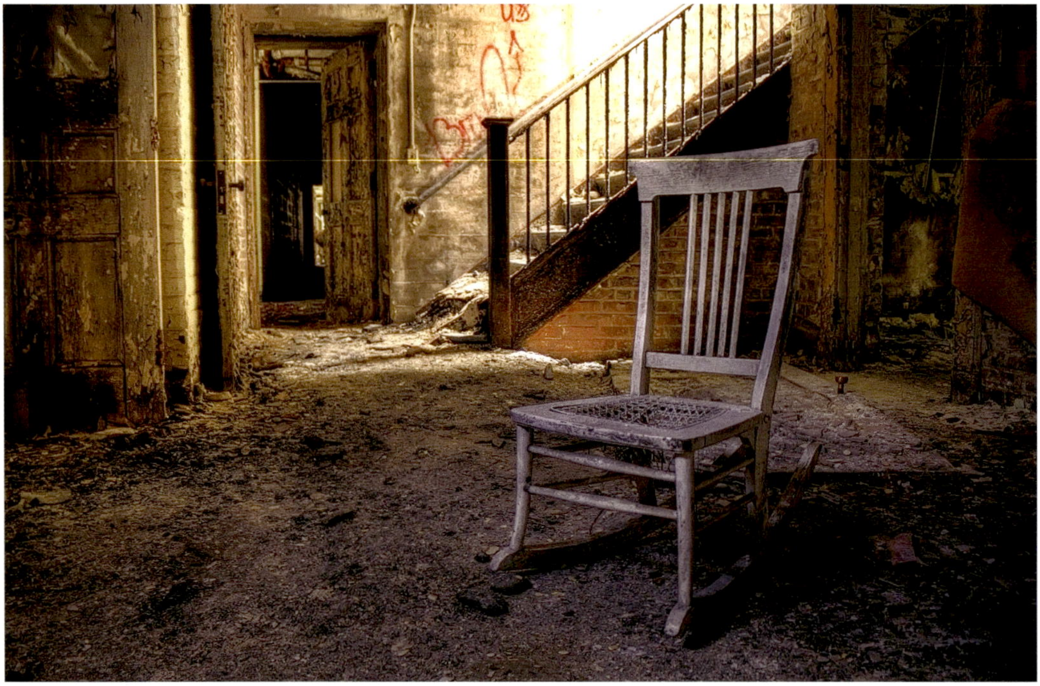

Explorers have a saying: "There's always a chair." This one is seen at Letchworth Village, an abandoned psychiatric institution in Thiells, New York. (*Photo: Andy Milford*)

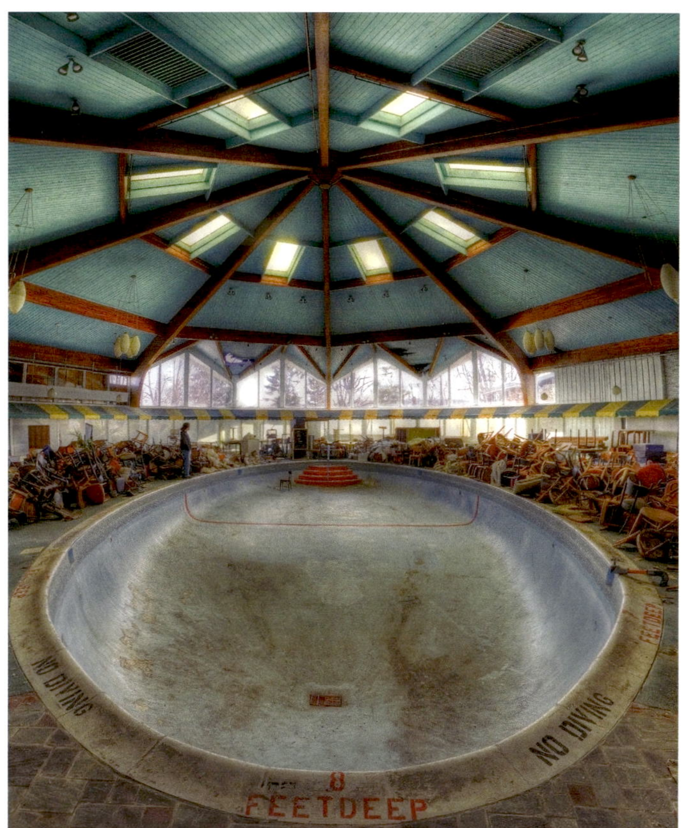

Right: Tamarack Lodge was one of many resorts in the Catskills known as the "Borscht Belt." A 2012 fire destroyed everything, including the pool house. (*Photo: Andy Milford*)

Below: After years as a tuberculosis sanitorium, this convalescent home in Dutchess County is awaiting a rebirth. (*Photo: Liz Cooke*)

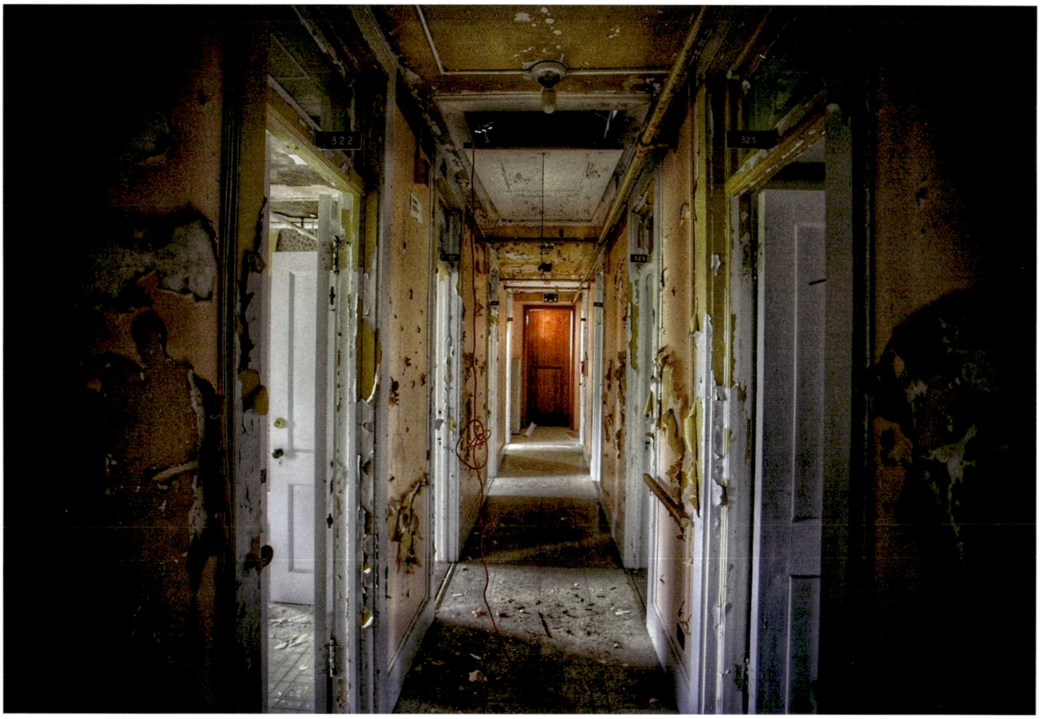

You will see examples of some of the finest architecture of the eighteenth, nineteenth, and twentieth centuries, eroding in real time.

People always ask what we will do when these places are gone, after all, many are imperiled due to neglect, others have demolition orders in place, and others are already gone. We reply that there will always be more. For some reason, the Hudson Valley has more than its fair share of abandoned places, providing end-less—and generally fruitful—inspiration for adventurous photographers. And while it is encouraging to see once-abandoned places brought back to life as arts centers and residential developments, we cannot help but be wistful for the stories lost to the wrecking ball.

When you see an old farmhouse with tattered curtains blowing in the wind or a historic mansion left vacant and forsaken, you might ask yourself why no one has come to its rescue. Or like us, you may take out a camera and capture what you see. To us, recording the moment is the best we can do. And if the picture moves someone, or stimulates some forgotten memory, all the better. The Hudson Valley has inspired countless artists and painters over the years. Some look at the majestic landscape, others look at the decaying farmhouse. Just so long as we keep looking, we will do justice to this singular, spectacular place.

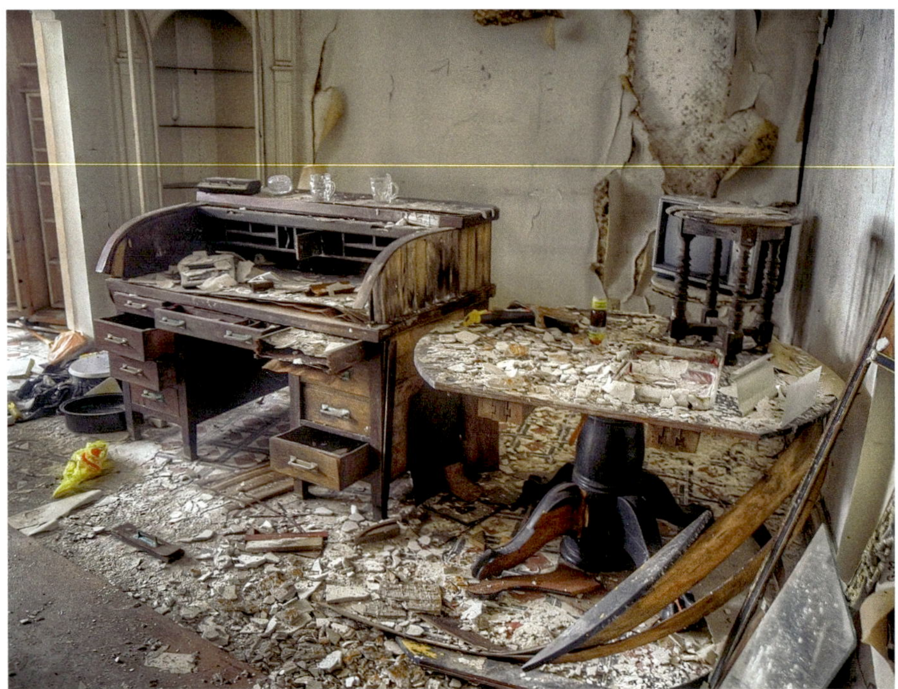

Scattered remnants of a life well lived are all that remain in this once-grand Catskills home. (*Photo: Andy Milford*)

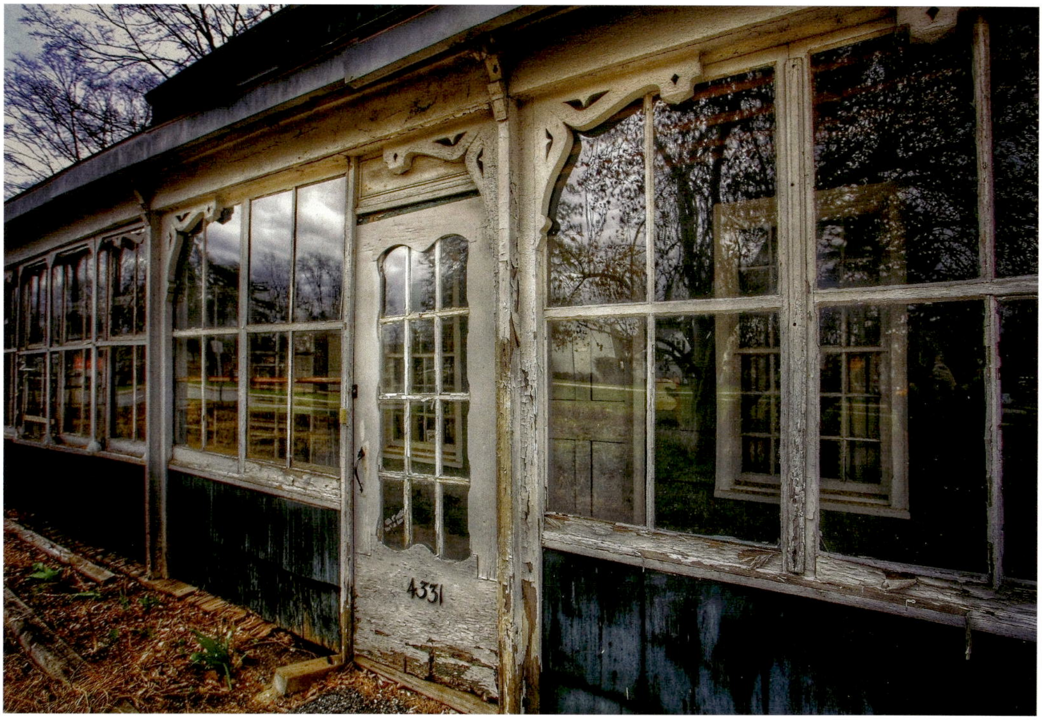

This modest home in Hyde Park, New York, boasts elegant details and local significance. Neighbors recall the beauty of the house and what a centerpiece it was "back in the day." (*Photo: Liz Cooke*)

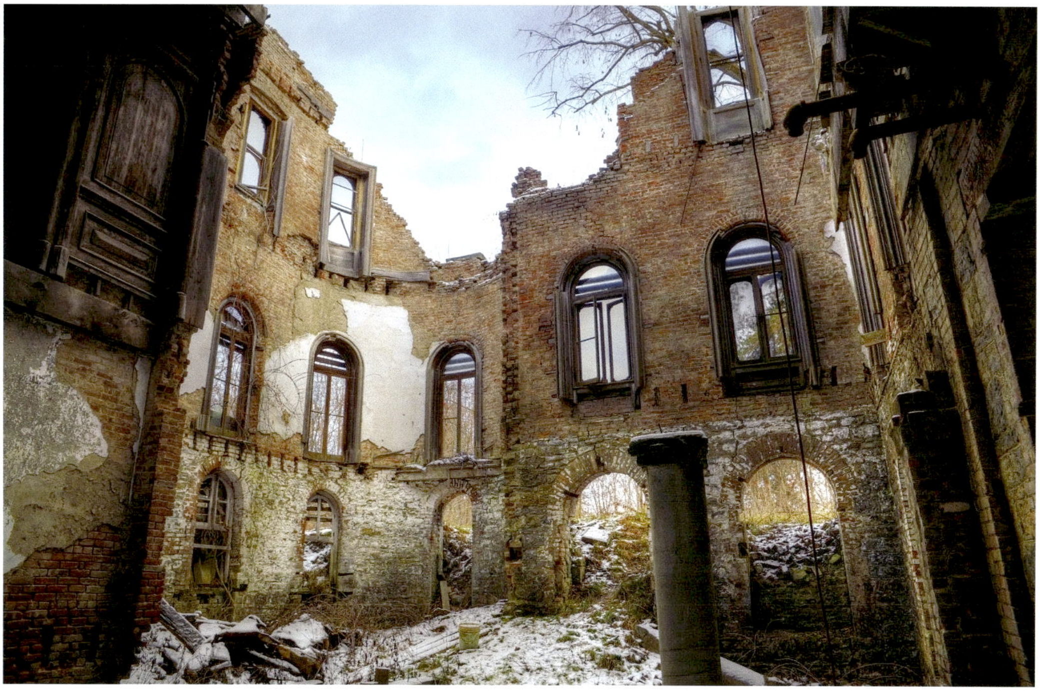

Interior walls in the crumbling Gilded Age mansion known as "Wyndcliffe" in Rhinebeck, New York. (*Photo: Andy Milford*)

1

HOMES, STONE HOUSES, AND FARMHOUSES

In any given area, there are any number of homes that can be called "abandoned." Not vacant—which could mean anything from an uninhabited vacation home to an unsold spec house to an unrented apartment—but abandoned, as in left behind, forgotten, and forsaken. Drive along any road in the Hudson Valley long enough and you are bound to see an abandoned home hidden behind overgrown bushes or in a stand of gnarled trees. Look further and you will likely spy one amid cornfields and apple orchards or even in residential cul-de-sacs and busy city centers.

Abandoned homes speak of melancholy, of unpaid taxes and reversals of fortune, of divorce and displacement, and sometimes—often—death. But look further and you will sense something else entirely. In the pages of a torn newspaper, you might read about a war hero or a high school graduation. You might see a wedding invitation or a dried bouquet of roses. There will be glassware and china, and the last bites of dinners at the table. Walking through an abandoned farmhouse, you can almost feel the presence of the families who once called it home. The rooms are empty, but traces of their lives remain—a baby carriage filled with dolls, a chair pulled up to a table as if someone had just gotten up from a meal. Outside, the fields lie fallow, the fences broken and overgrown. The barn, once a hive of activity, now stands empty, its stalls deserted and its hayloft silent.

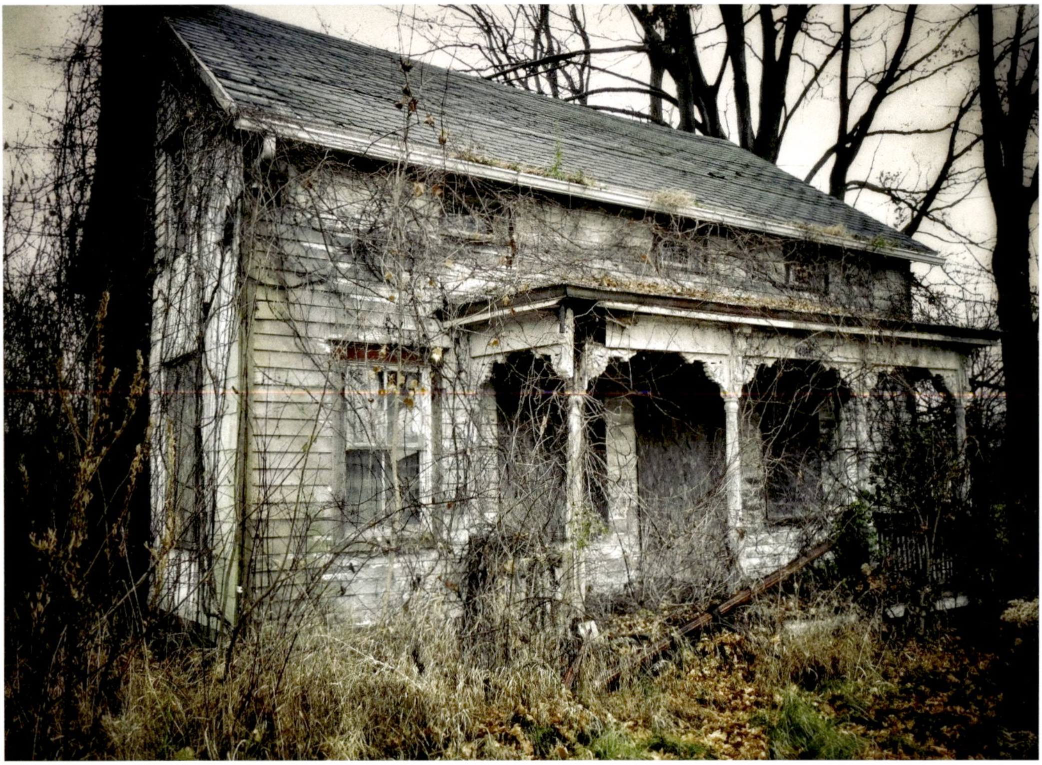

The "Vine House" in Bells Pond, New York, looks like something out of a Stephen King novel. Location scouts and local photographers share a fascination with this intriguing structure. (*Photo: Liz Cooke*)

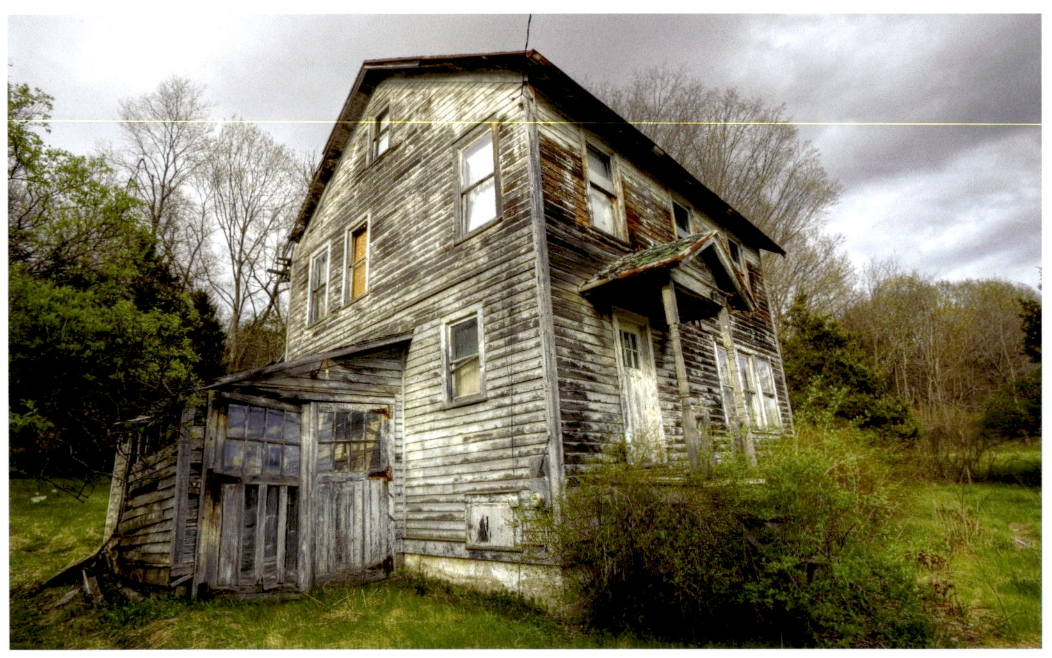

Abandoned farmhouses abound in the Hudson Valley, perfect set pieces for curious photographers. This one is known as the "Forgotten Farmhouse." (*Photo: Andy Milford*)

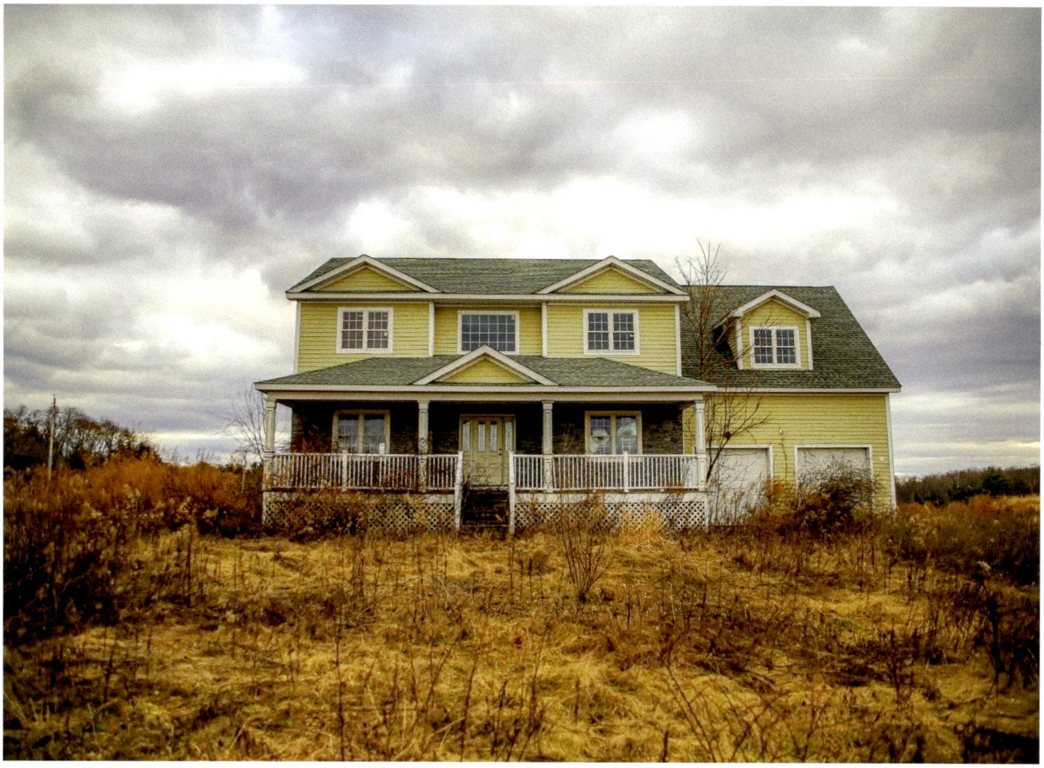

This upscale contemporary home lies vacant in a Dutchess County field. For reasons that are unknown it has never been occupied. (*Photo: Liz Cooke*)

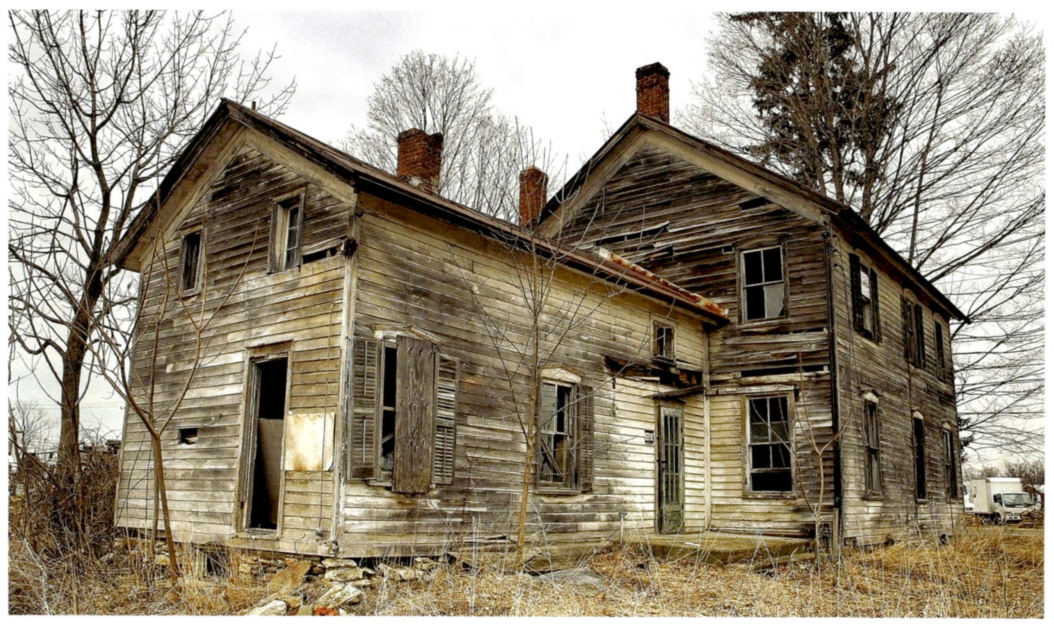

Why are we so enamored with the beauty of a decaying farmhouse? Perhaps we long for a simpler time, when food and family were all that mattered? This abandoned Columbia County farmhouse checks all the boxes. (*Photo: Liz Cooke*)

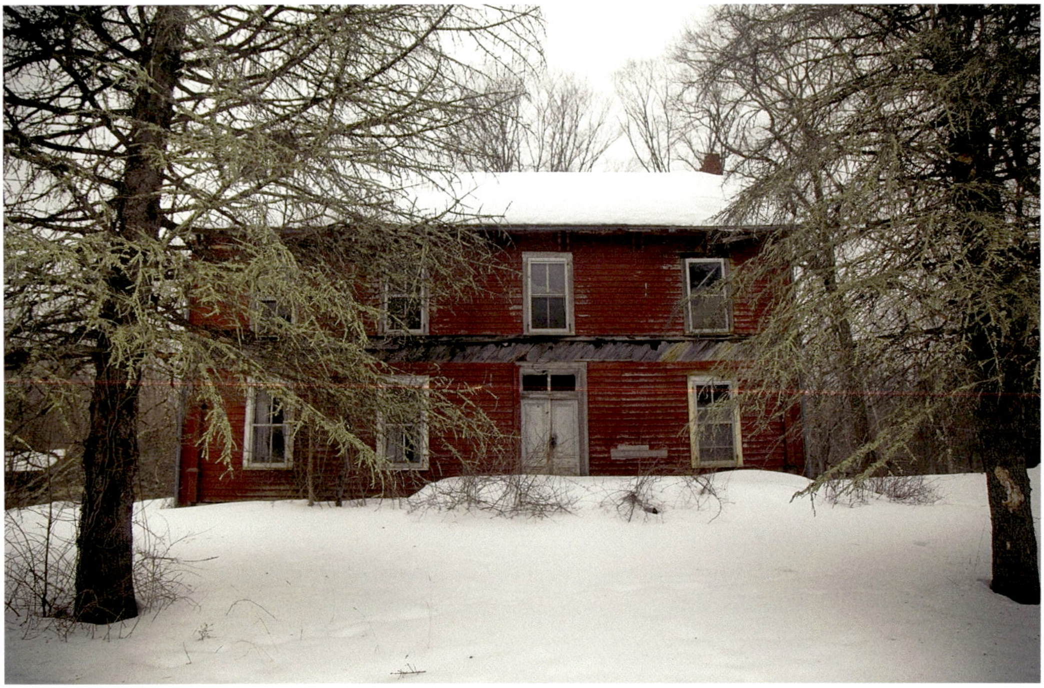

Even in the depths of a snowy Hudson Valley winter, the beauty of an abandoned red house shines though. (*Photo: Andy Milford*)

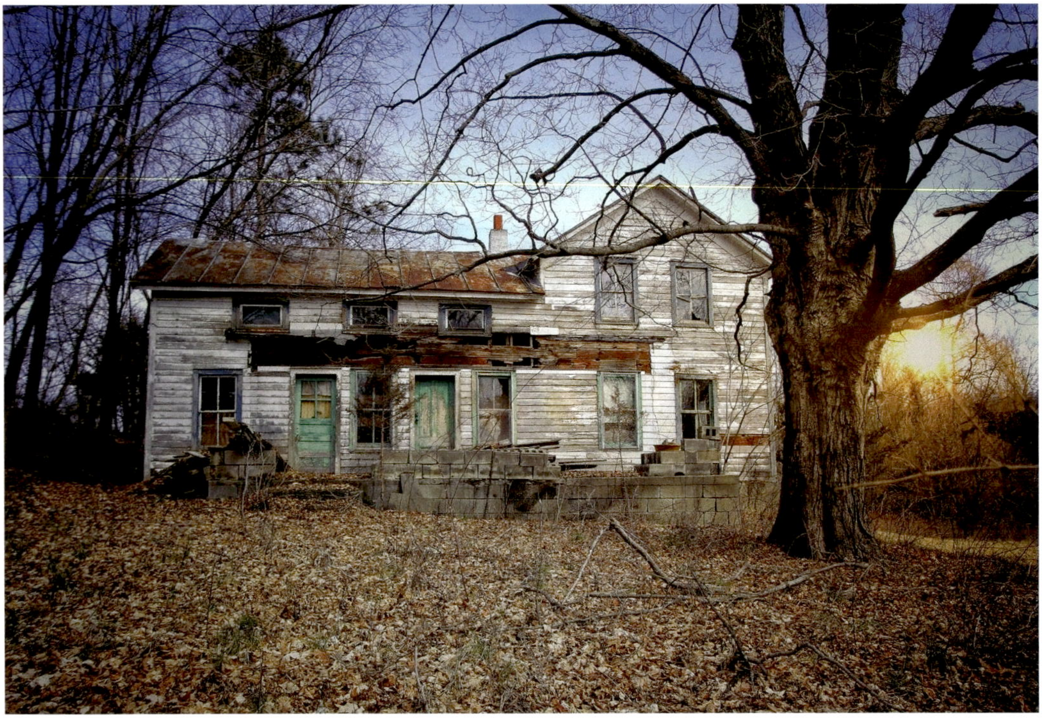

The sun is setting on this abandoned home we call the "Green Door House." We were intrigued by the tiny second-story windows. (*Photo: Liz Cooke*)

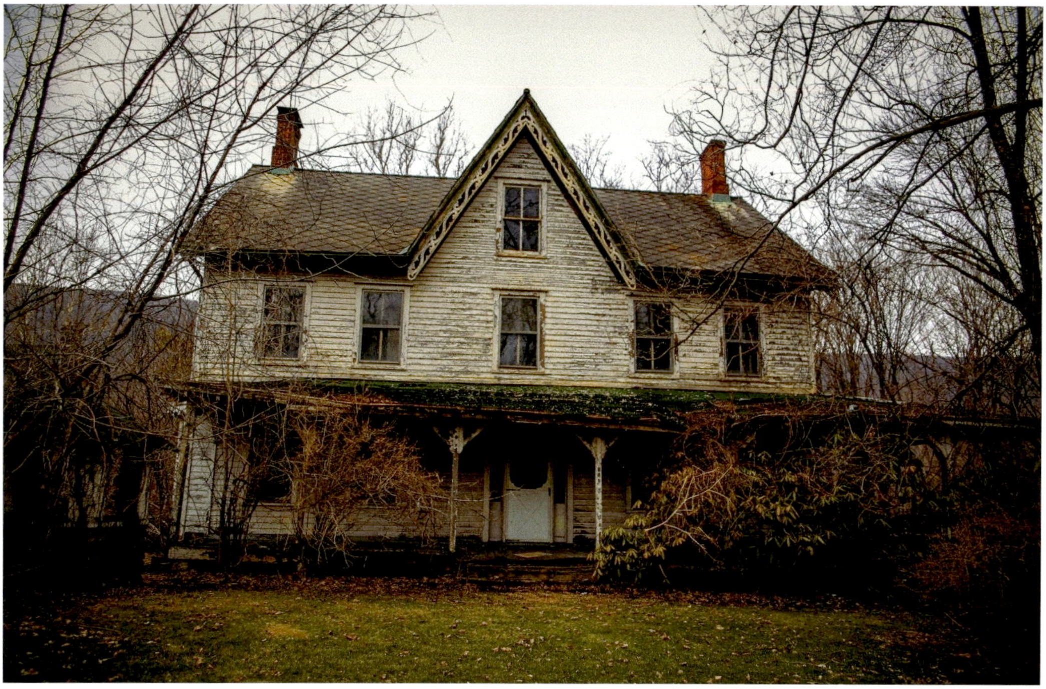

One can only imagine the lives lived in this abandoned Orange County beauty. (*Photo: Liz Cooke*)

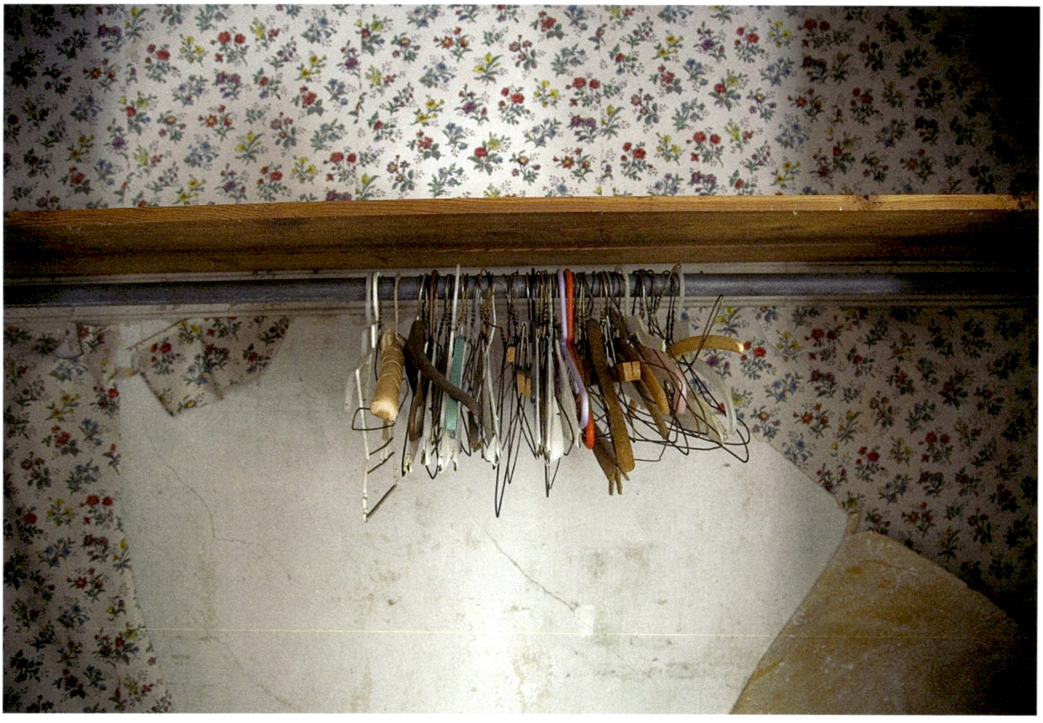

A collection of hangers is all that remains in the wallpapered closet of an abandoned farmhouse. (*Photo: Andy Milford*)

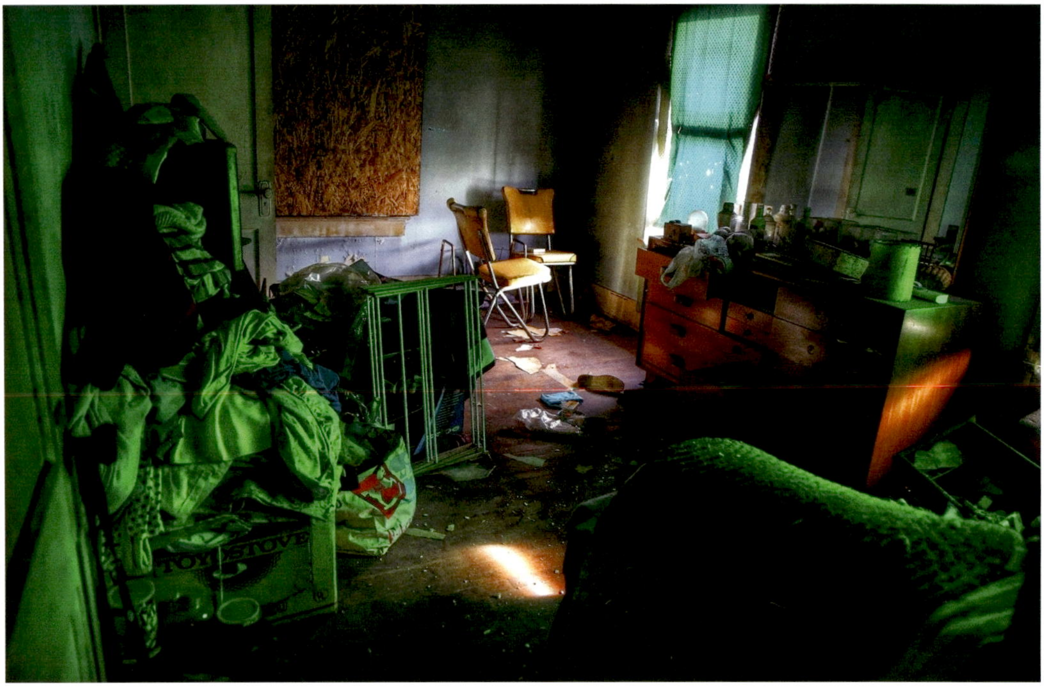

Set on a busy road in Hudson, New York, this old house casts a surreal green light on a cluttered room. But with two inviting chairs, who wouldn't want to linger? (*Photo: Liz Cooke*)

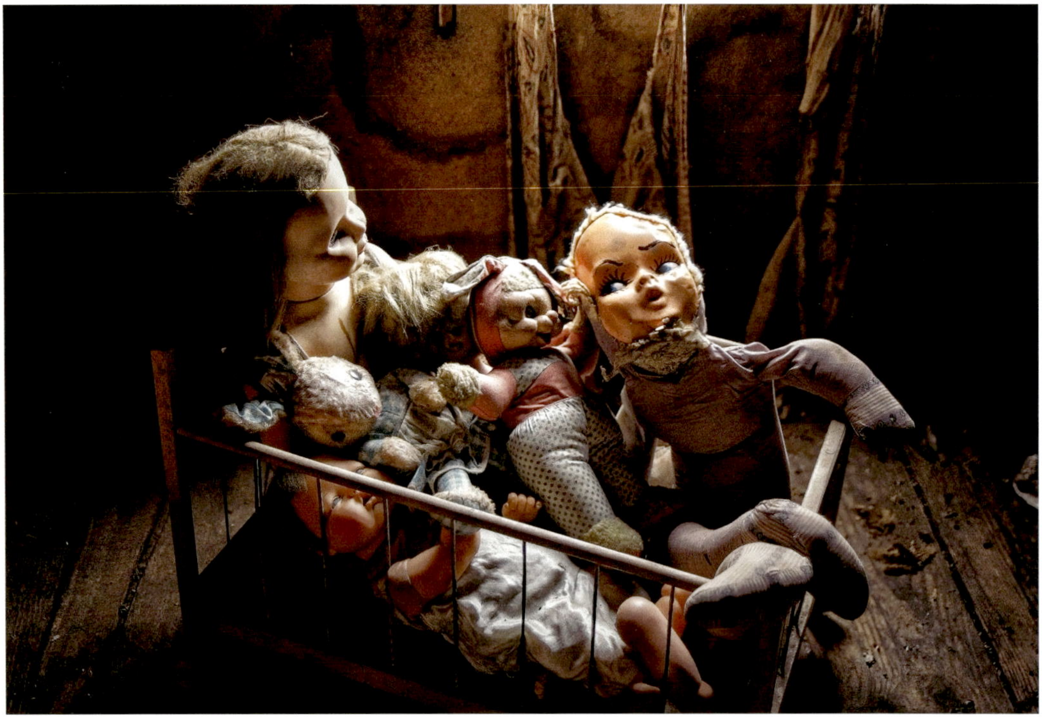

A child's crib filled with dolls was discovered in an attic in an abandoned Hudson, New York, home. (*Photo: Liz Cooke*)

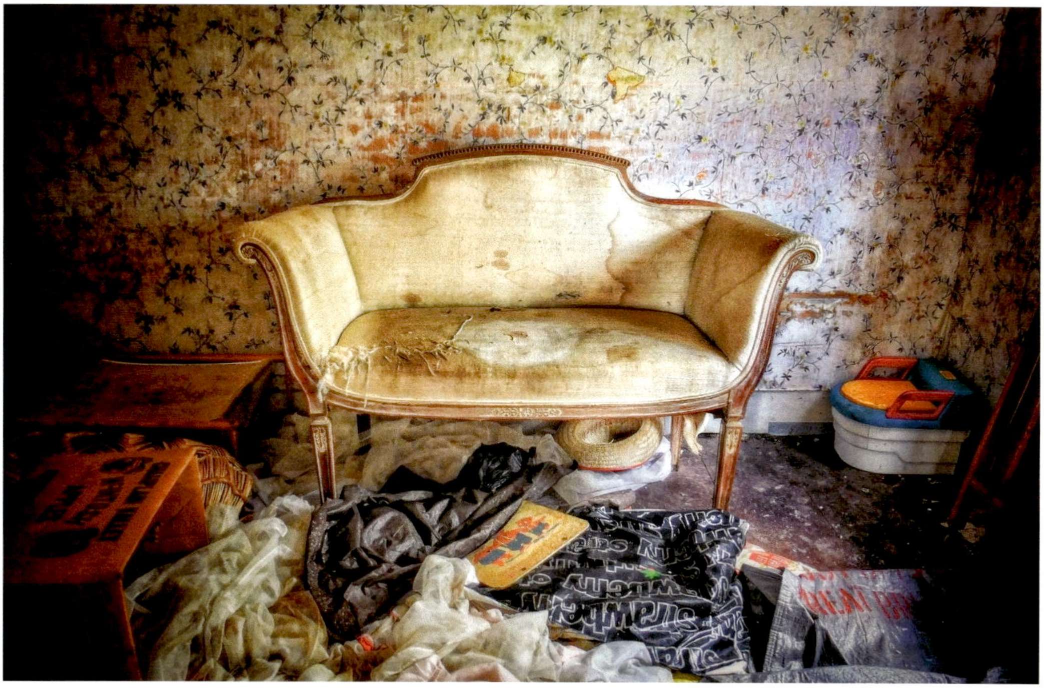

A silk divan and a child's potty, seen in an abandoned Greene County home. (*Photo: Liz Cooke*)

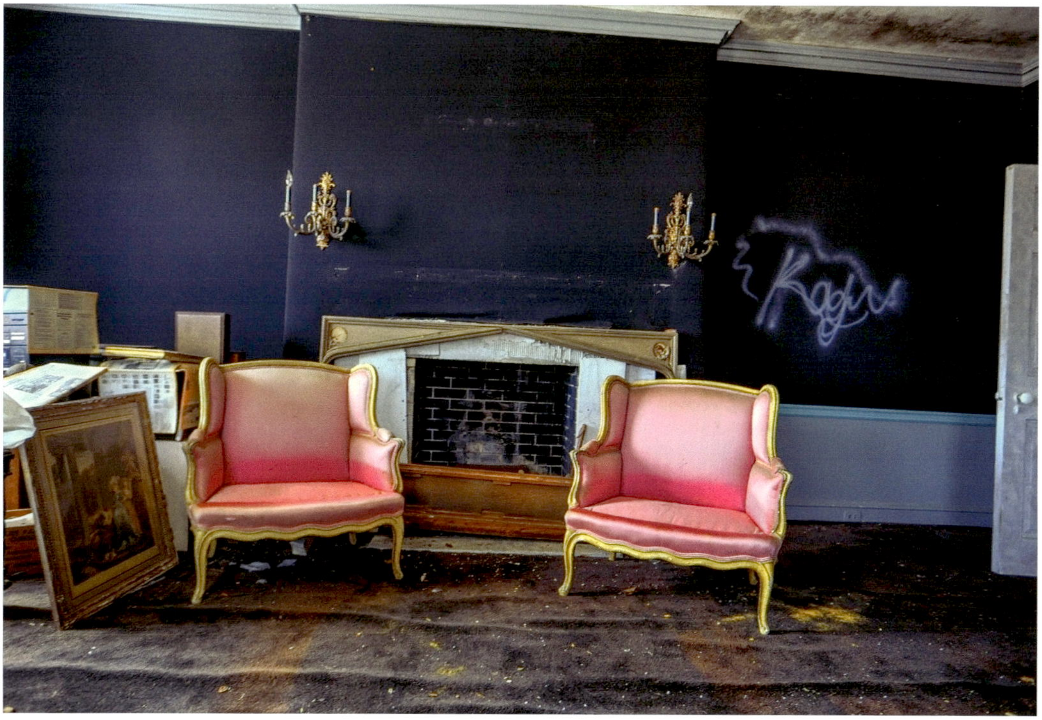

Two pink chairs, elegantly arranged, seen in an abandoned home we call the "Musty Mansion." (*Photo: Andy Milford*)

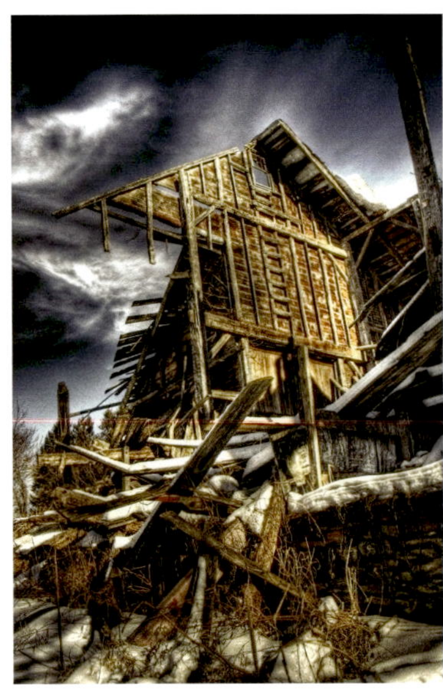

We can only assume the "Stormy Barn" was once a bustling hub of activity on this abandoned family farm in Rhinebeck, New York. (*Photo: Andy Milford*)

Besides rustic farmhouses, the Hudson Valley boasts many stone houses built by Dutch settlers in the seventeenth and eighteenth centuries. Sadly, many of these historic structures are fully abandoned, deteriorating in plain sight, and in some cases, covered in graffiti. One abandoned (or at least vacant and unused) stone house is located in Poughkeepsie, New York. The Abraham Fort Homestead was built in 1743 and played a prominent role in the American Revolution. Over the years, ownership has changed many times. At one point, it was used as guest accommodations for visiting IBM executives. Today it sits vacant (although well preserved) just down the road from auto supply stores, budget hotels, and a Buffalo Wild Wings.

Some houses meet dire fates despite their historical significance. The Jackson House in Fishkill, New York, was a significant historic landmark with a storied colonial past. Constructed in 1741, the house was one of the oldest in the area. Despite efforts from community organizers and preservationists, Jackson House fell into disrepair and neglect, leading to its demolition in 2013.

While abandoned stone houses, historic houses, and farmhouses can evoke romantic images, that is not the case for every house. One of the eeriest and most tragic abandoned homes in the region belonged to convicted serial killer Nathaniel White. Located in Goshen, New York, this unassuming white house would be used by the killer to carry out his horrific crimes. Following his capture, the notoriety of the house led to it becoming a focal point of fear and curiosity. Over time, the property fell into further disrepair and was eventually demolished.

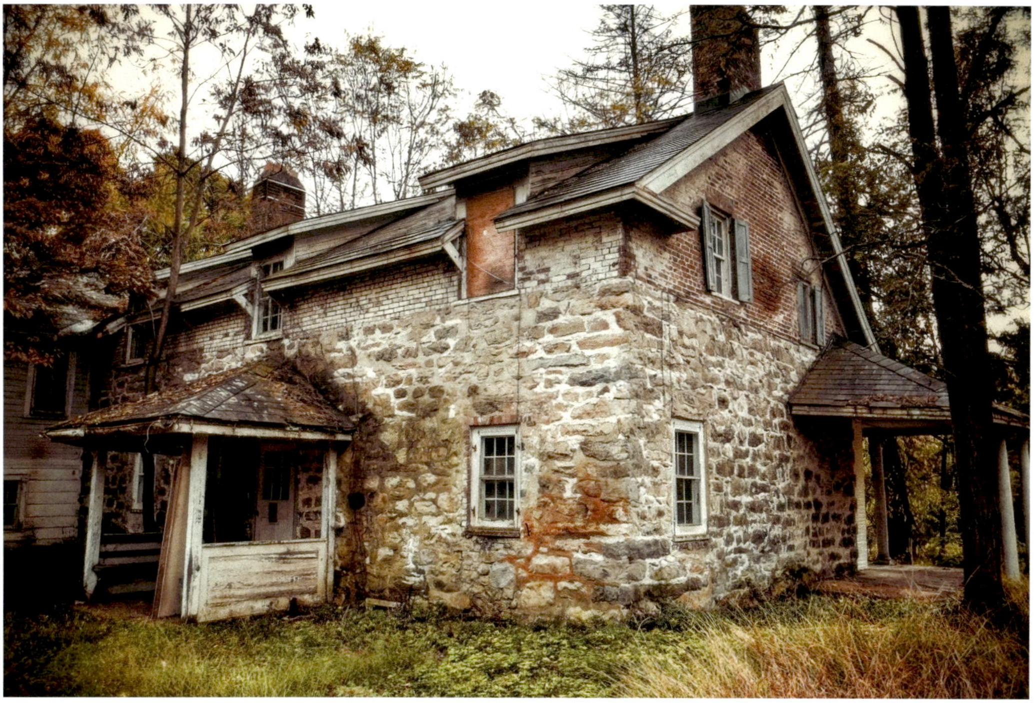

The Abraham Fort House (also called the Fort Homestead), located in Poughkeepsie, New York, had a long Revolutionary War history but was abandoned in the mid-twentieth century. (*Photo: Liz Cooke*)

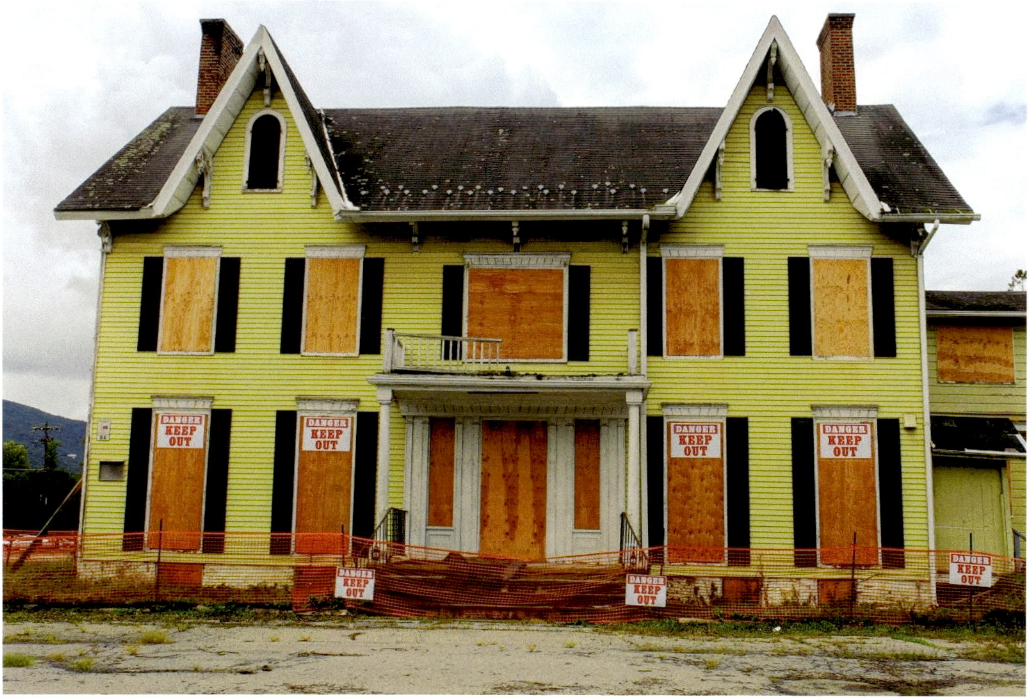

Despite the efforts of local preservationists, the historic Jackson House in Fishkill, New York, was demolished in 2013. (*Photo: Liz Cooke*)

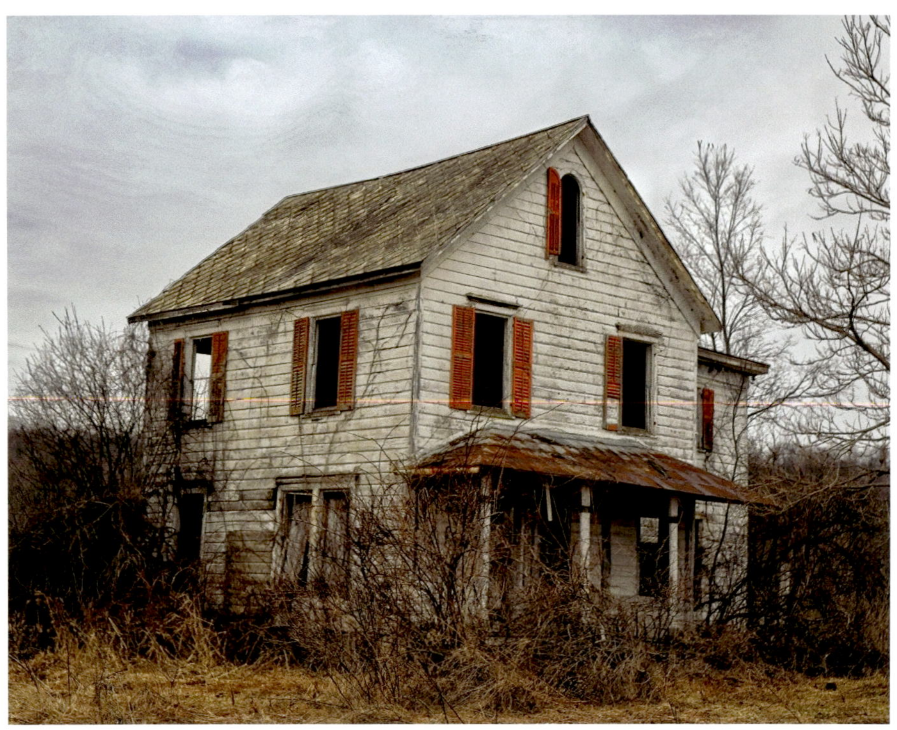

This unassuming house in Goshen, New York, was the scene of gruesome crimes committed by notorious killer Nathaniel White. (*Photo: Liz Cooke*)

Other houses we explored include the "Mold House," a contemporary home set amid an idyllic countryside backdrop. What sets this house apart from its inhabited neighbors has to be the thick layer of mold that covers almost every surface. Fascinating though it was, we did not spend long here. We also explored a historic home we call the "Sea Shanty." While not situated on the sea, this home does show evidence of a seafaring life. With such an abundance of art and artifacts, it was hard to know where to look. Similarly, we visited an abandoned house we call the "Arts Space." Located in the deep Catskills, this house boasted one of the most unusual collections we had ever seen—a room filled top to bottom with pink flannel onesies. And finally, there is the "Time Capsule" house, an abandoned home filled with memorabilia.

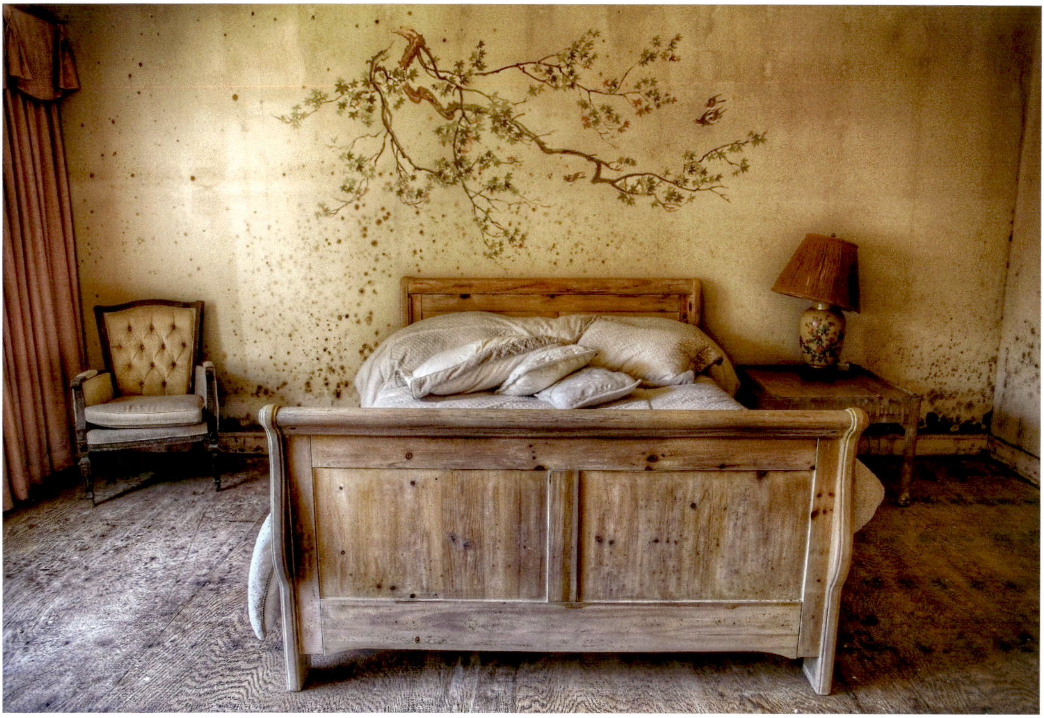

Located in a remote section of the Catskills, the "Mold House" was once a contemporary "dream home."
(*Photo: Liz Cooke*)

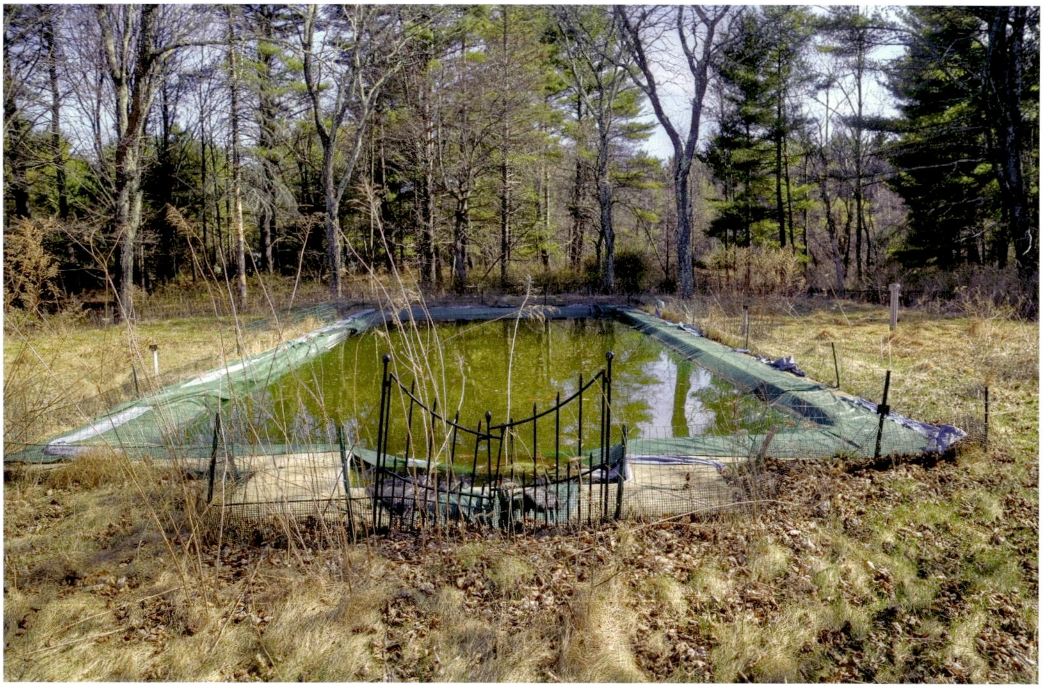

The grounds of the "Mold House" reveal an unused swimming pool and the vestiges of landscaping.
(*Photo: Andy Milford*)

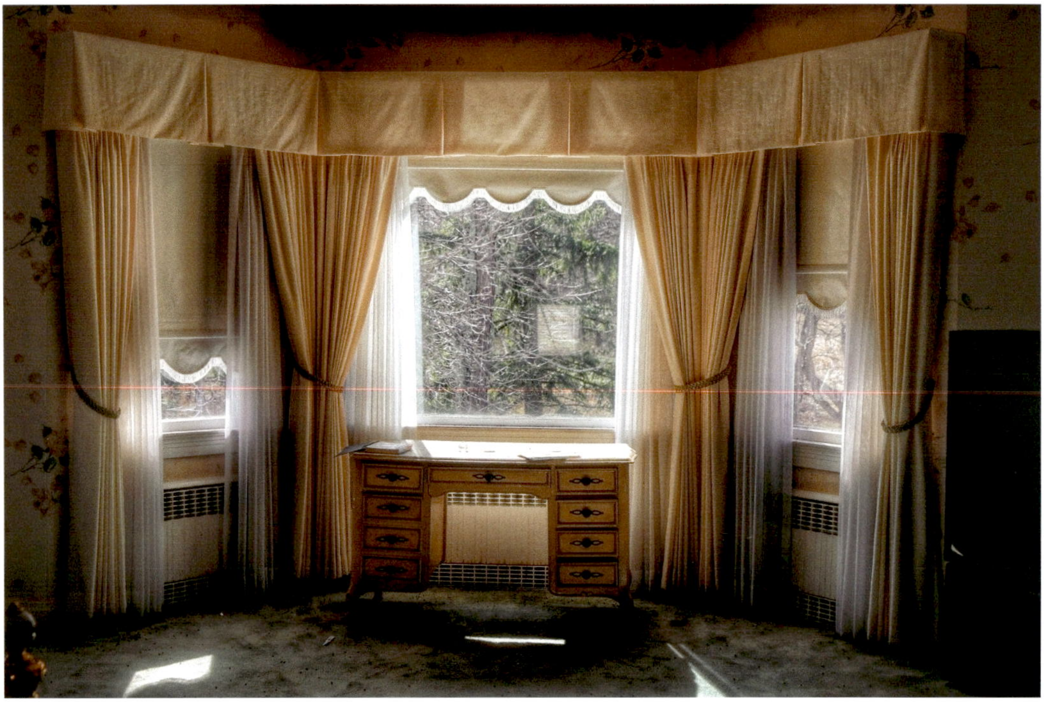

Draperies and fine furnishings are found in every room of the abandoned "Mold House." (*Photo: Liz Cooke*)

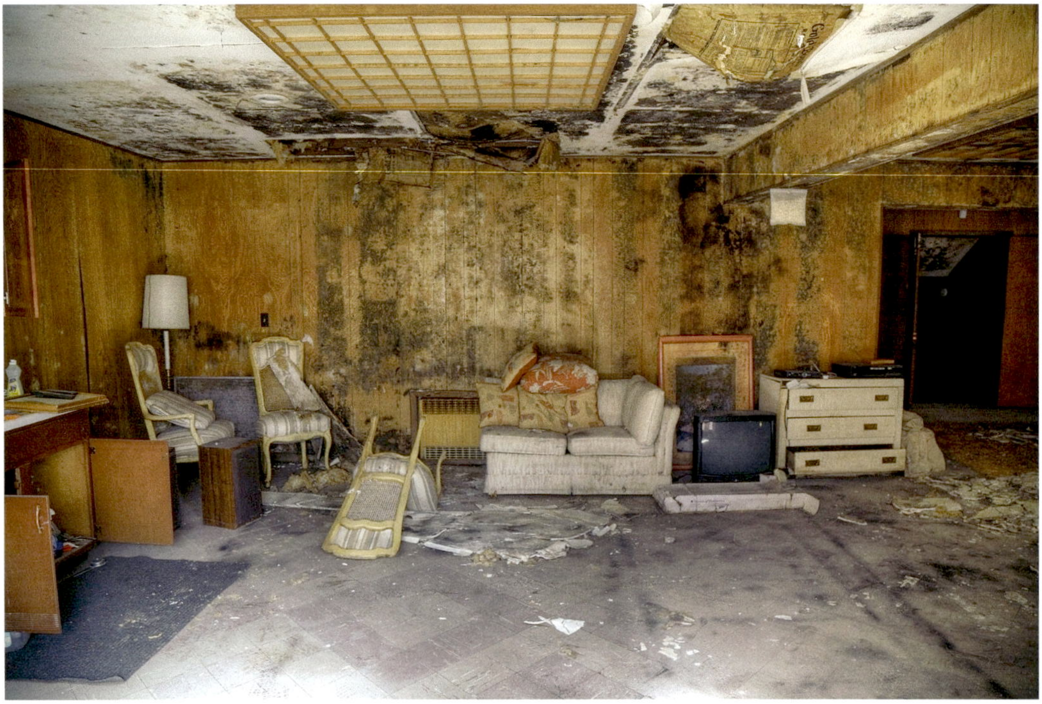

A thick layer of mold covers every inch of the home and furnishings in this Catskill Mountain abandonment. (*Photo: Andy Milford*)

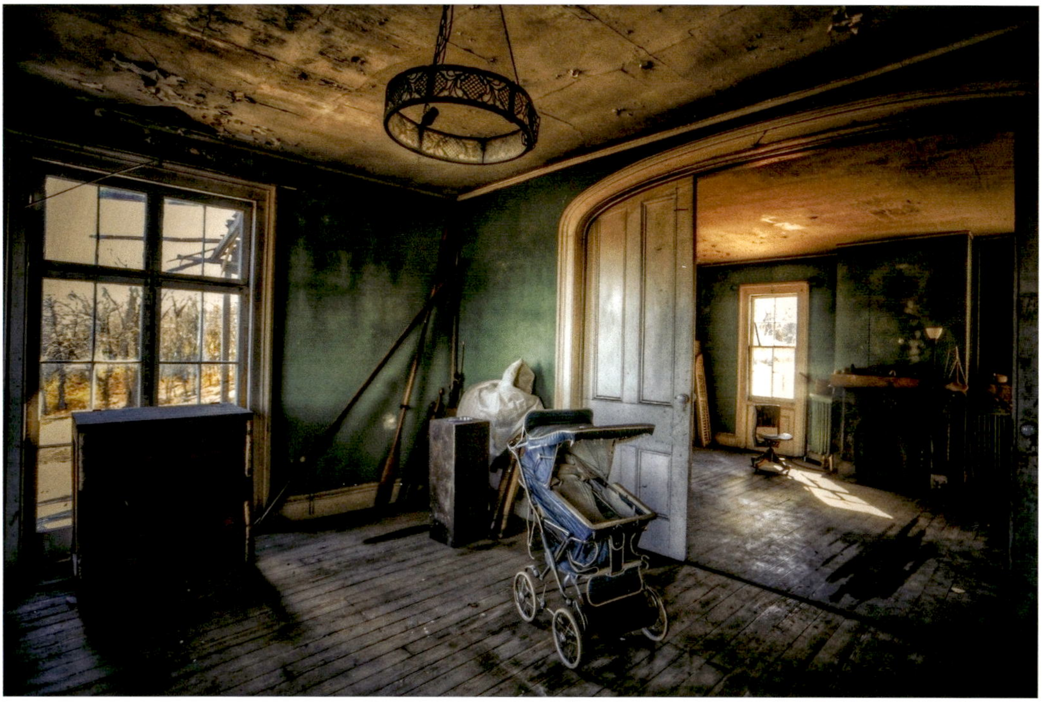

Offering a commanding view of the Hudson River, this abandoned house had several rooms full of intriguing artifacts. (*Photo: Liz Cooke*)

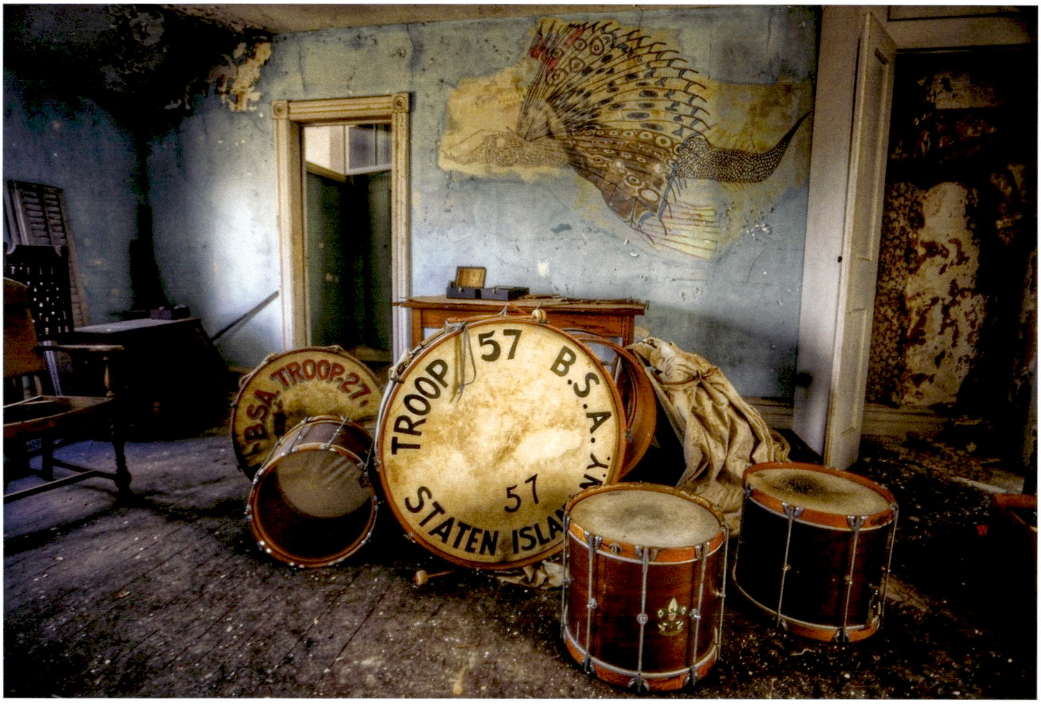

The history of this drum set, found in an abandoned home perched atop the Hudson River, is destined to remain a mystery. (*Photo: Liz Cooke*)

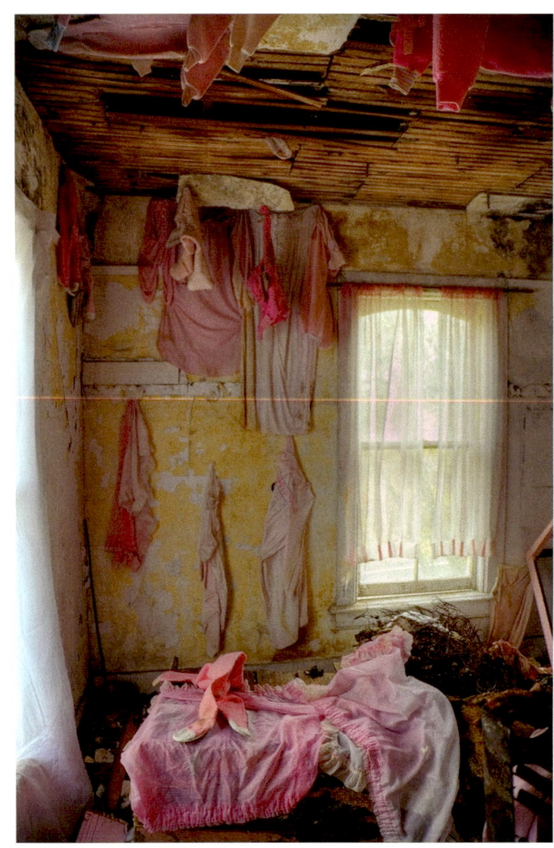

Left: Located near the hamlet of Lexington, New York, this abandoned gallery space hosted an exhibition of onesies covering an entire room. (*Photo: Andy Milford*)

Below: This paint-splattered room was likely part of an artist's show at a Lexington, New York, abandoned gallery space. (*Photo: Liz Cooke*)

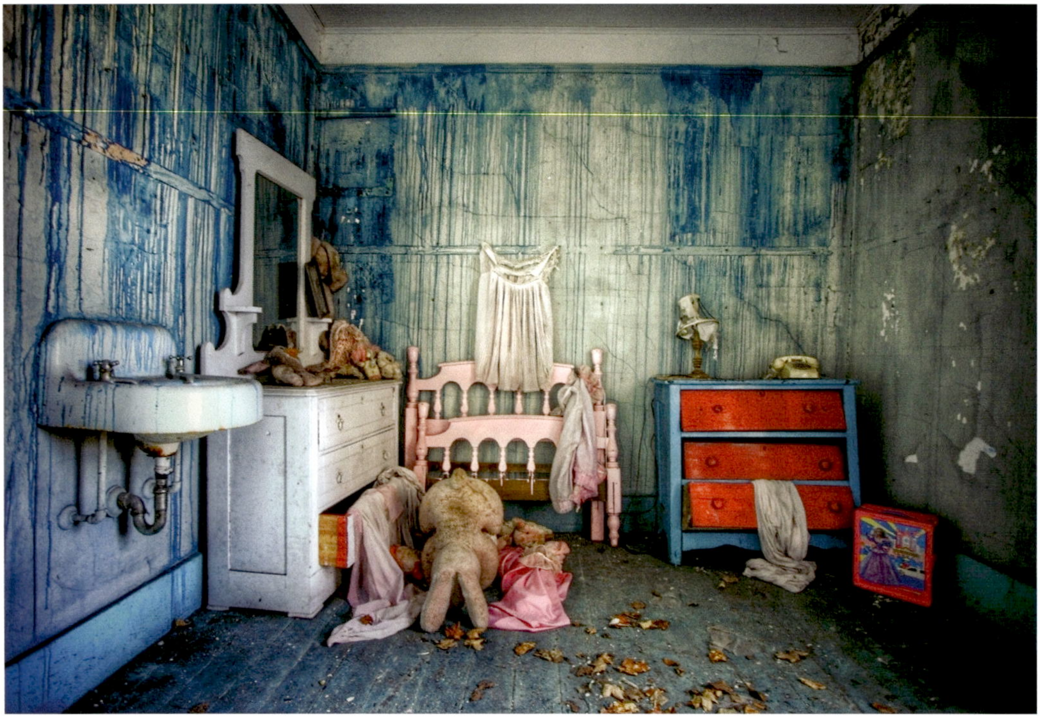

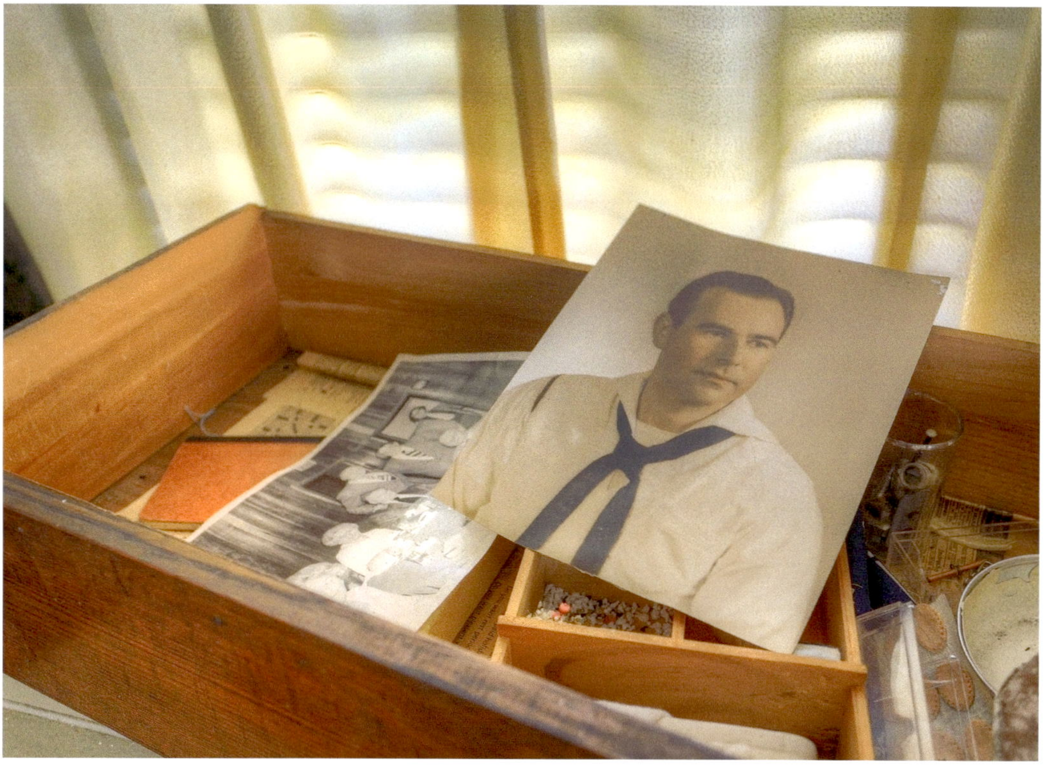

This photograph of a sailor was just one personal artifact left behind in a Columbia County abandonment. (*Photo: Andy Milford*)

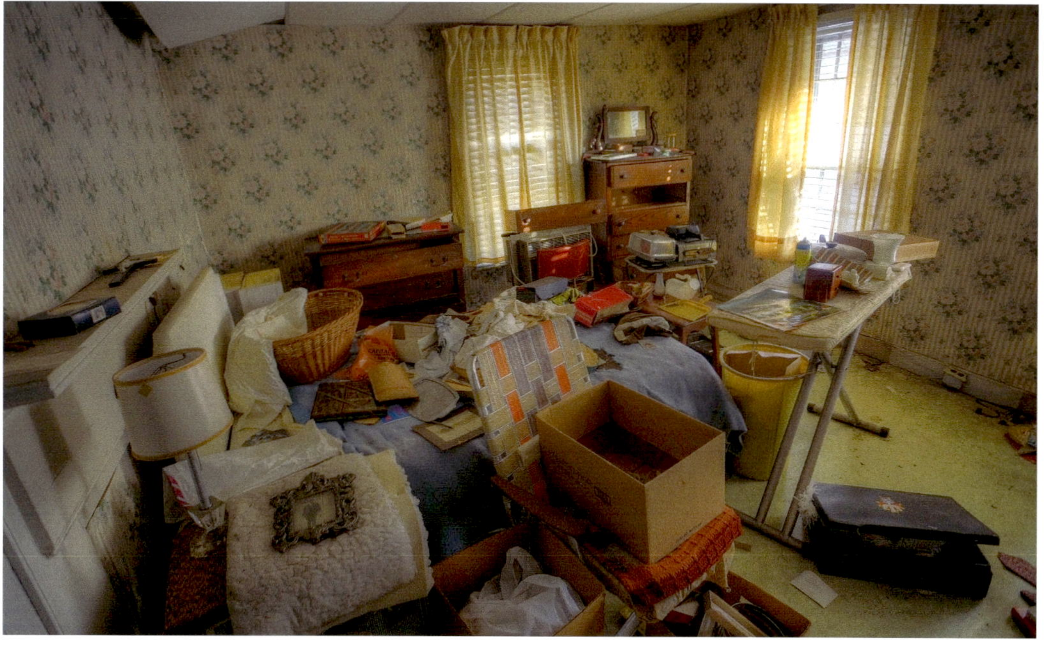

A profusion of personal items, some in pristine condition, suggests a hasty departure from this home near the Hudson River in Columbia County. (*Photo: Andy Milford*)

2

MANSIONS AND ESTATES

One of the most compelling features of the Hudson Valley is the sheer number of grand mansions situated on the shores of the Hudson River and throughout the countryside. Many of these estates were built during the Gilded Age, a period of immense wealth and opulence in the late nineteenth and early twentieth centuries. Members of New York City's aristocracy would decamp to their summer dwellings looking for respite from the sweltering heat of the city. They did not, however, "go country" when they came north. Instead, they built homes of such size and scope that they rivaled even the grandest mansions of Fifth Avenue. Enormous rooms festooned with brocade curtains, silk tapestries, and priceless antiques and artifacts—many shipped by boat from Europe and Asia—decorated these "country cottages."

Life in the Hudson River Valley was fine indeed. Lawn parties, literary salons, and lazy hours lolling with friends and family filled their days. Worries about wars overseas and poverty at home were the farthest things from their minds. For some, it seemed like wealth and privilege would protect them from the harsher realities of life. It would not, however, turn out that way. Through death, divorce, inheritance issues, and high taxes, many could not sustain this opulent lifestyle and simply walked away. Others tried to hold on, only to ultimately be forced to move on. For those, the death knell would be the need for modernization, whether it was to update plumbing to modern codes or adding sufficient electric power for a contemporary lifestyle. All these expenses were too costly for families whose "glory days" were well in the past.

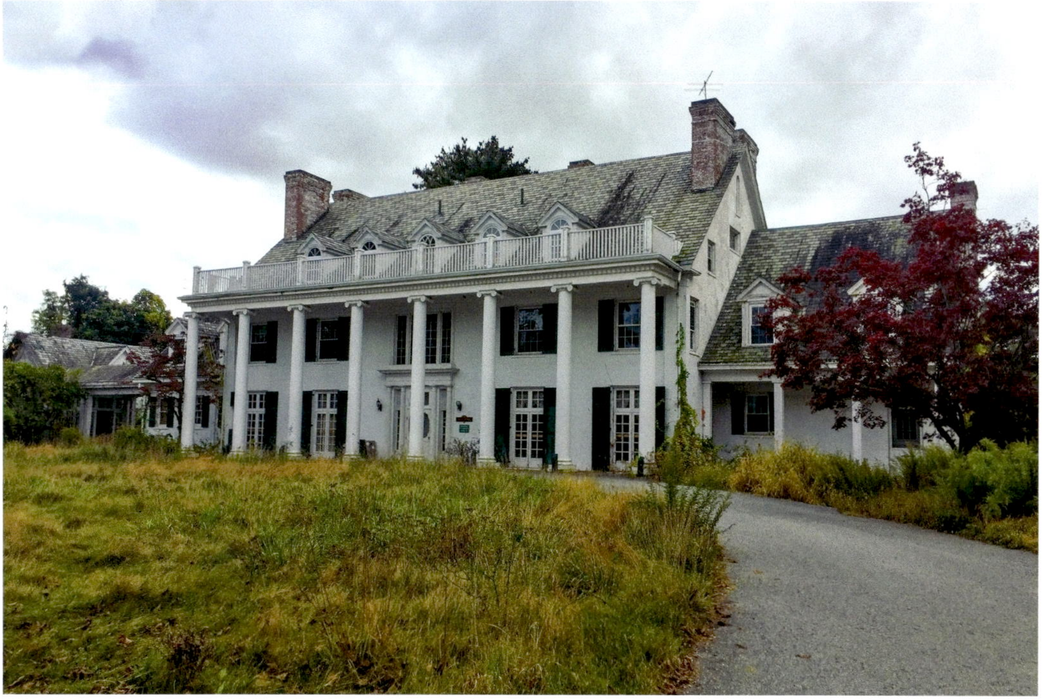

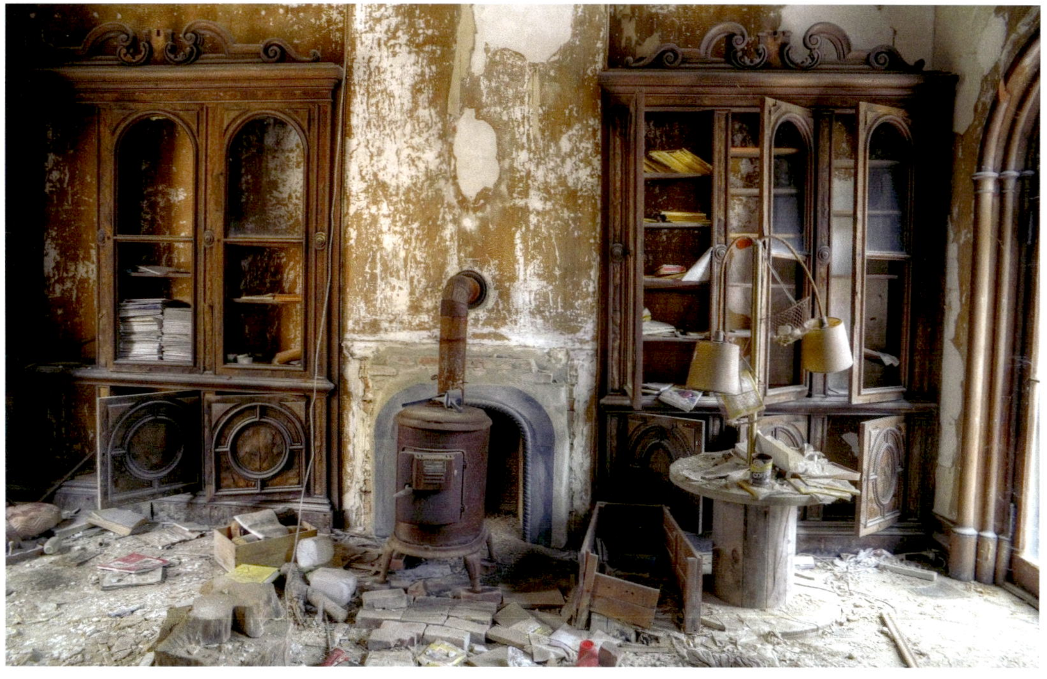

Known as the Tracy Dows estate or Foxhollow Farm, this Rhinebeck, New York, mansion was once a prestigious private school and later used for storage by a substance abuse facility. (*Photo: Liz Cooke*)

Master craftsmanship was a hallmark of the Rhinebeck, New York, mansion known as Finck Castle, Linden Hall, and now "Wyndcliffe." Abandoned for decades (but still standing), it is undergoing complete restoration. (*Photo: Andy Milford*)

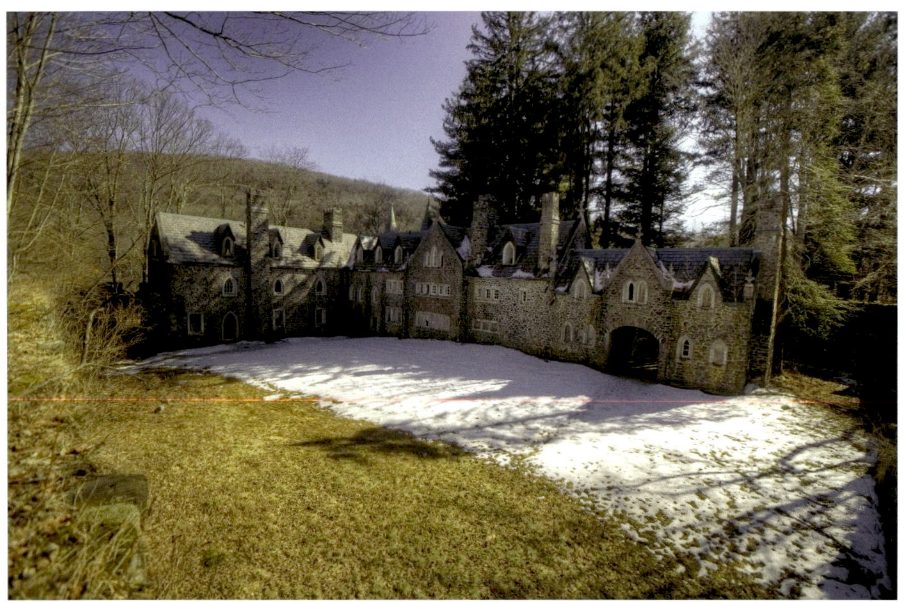

Located near Roscoe, New York, Craig-E-Claire Castle or Dundas Castle was built as an expression of love for the owner's wife and never occupied by her. It was later used as a summer camp and a Masons' lodge before becoming completely abandoned and vandalized. (*Photo: Andy Milford*)

One Hudson River mansion is Wyndcliffe, located in Rhinebeck, New York. Built in 1853, it was once a stunning example of Victorian Gothic architecture. The home of Elizabeth Schermerhorn Jones, Wyndcliffe is said to have inspired the phrase "keeping up with the Joneses." Over the years ownership changed many times, and the mansion was abandoned in the 1950s. With no caretaking it deteriorated quickly, its exquisite brickwork and fine interiors crumbling in plain sight. As of 2024, work has begun to stabilize the structure and perhaps return it to its former (although modernized) glory.

Another abandoned mansion is Hoyt House ("The Point") in Staatsburg, New York. Designed in 1855 by Calvert Vaux, the architect best known for co-designing Central Park with partner Frederick Law Olmsted (the team also designed the grounds of Hudson River State Hospital), Hoyt House was home to several members of the Livingston family. Despite efforts to keep the home in the Hoyt family, in 1962 the New York State Office of Parks and Recreation purchased the property to extend a stretch of riverfront parkland. The plan was to raze the house (which fortunately never happened), and the state authorities seemingly walked away from the project. Without maintenance, Hoyt House fell into disrepair and was threatened with the wrecking ball many times until preservationists united to save it. Much work remains to be done, and it is unclear what the future holds for this once-grand estate.

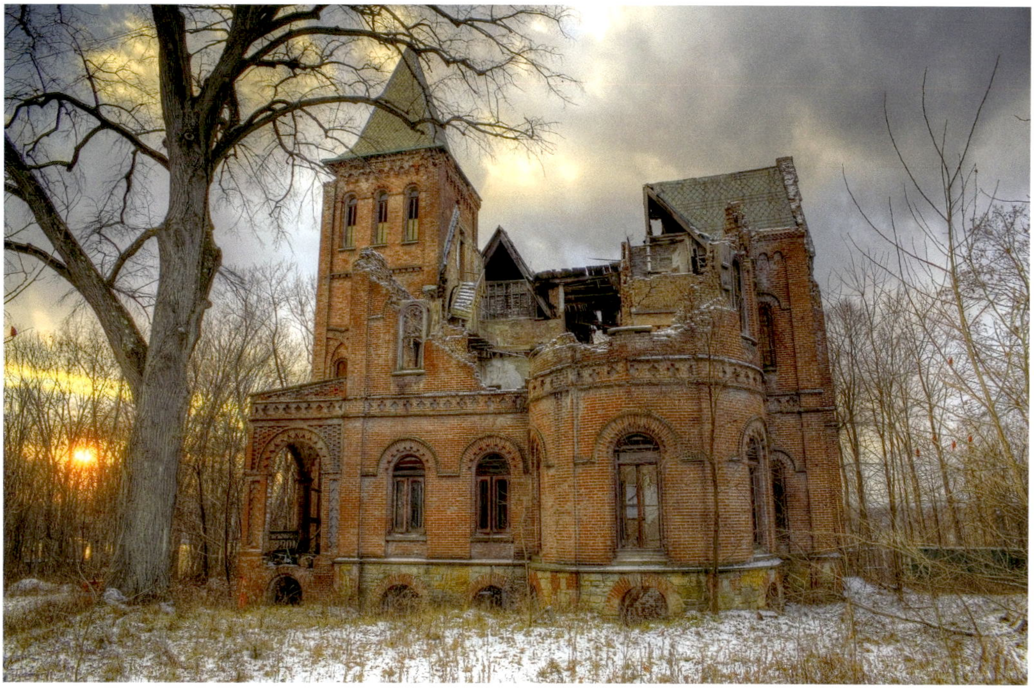

A picturesque moment at the Wyndcliffe mansion in Rhinebeck, New York, harkens to a time when New York's literary elite, including Edith Wharton, were guests here. (*Photo: Andy Milford*)

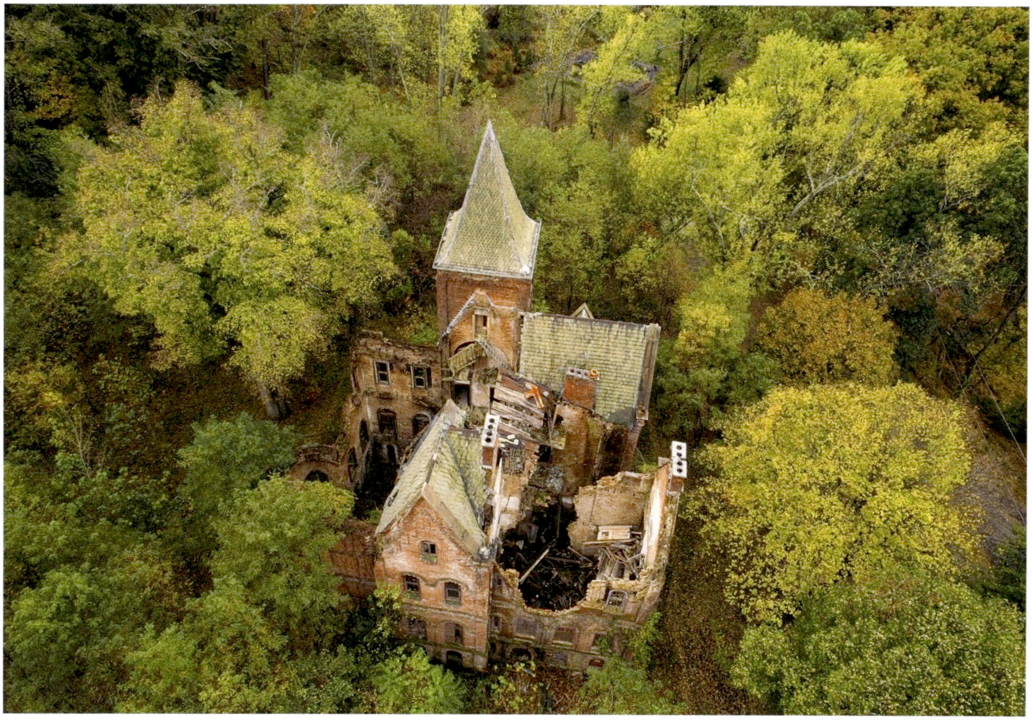

An aerial view of the Rhinebeck, New York, property known as "Wyndcliffe" clearly shows the ravages of time and neglect. (*Photo: Andy Milford*)

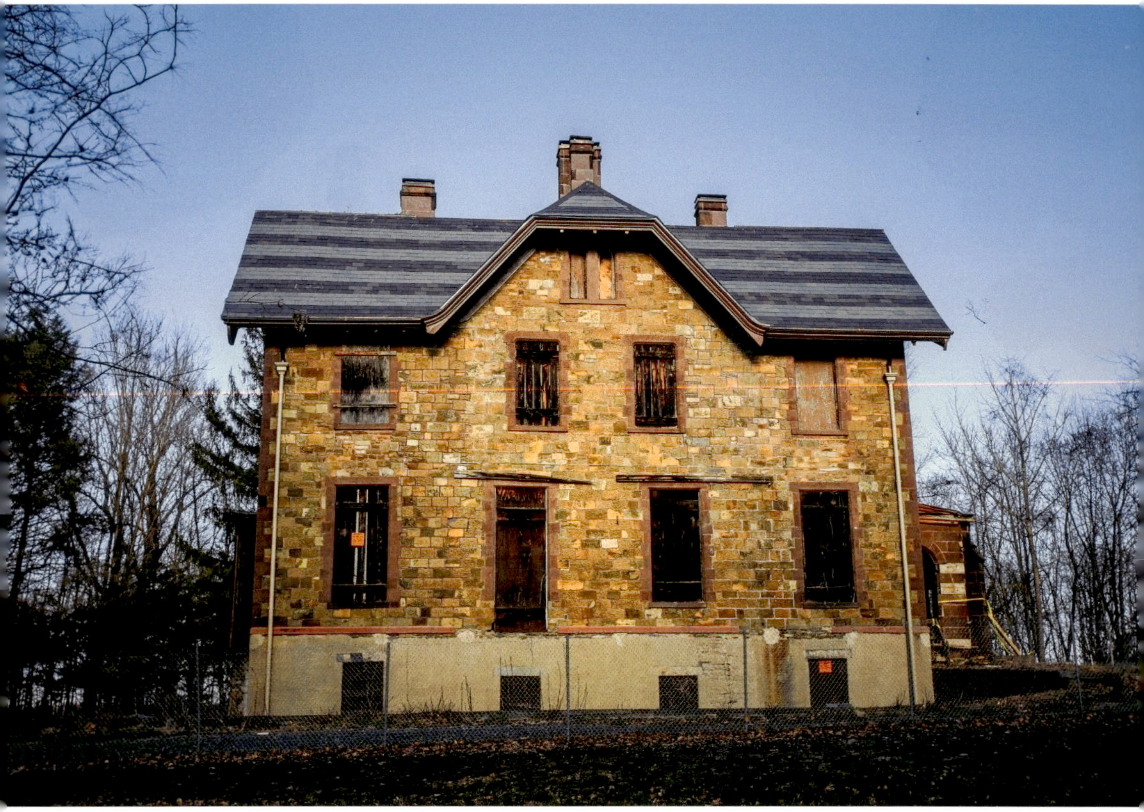

Located on the grounds of Mills-Norrie State Park in Staatsburg, New York, is the abandoned Hoyt House mansion. Despite efforts by local preservationists to find an adaptive reuse for the property, it remains abandoned. (*Photo: Liz Cooke*)

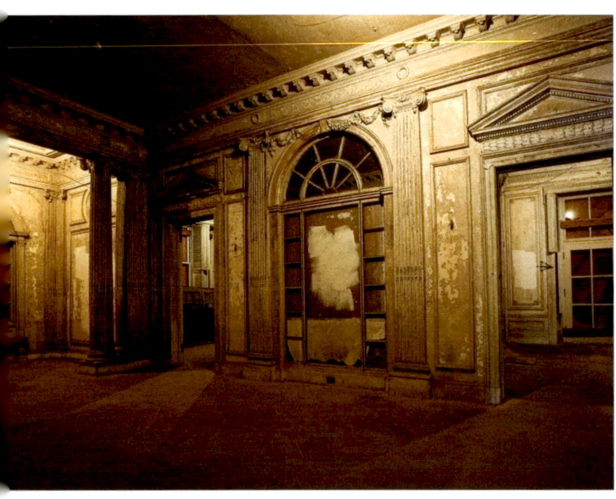

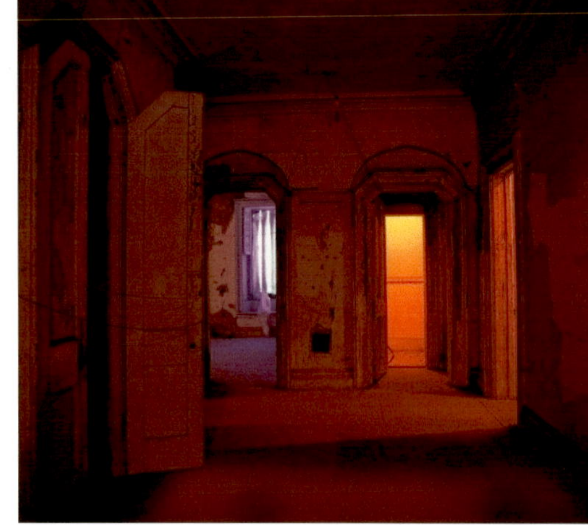

Above left: An interior shot of Hoyt House (also known as "The Point") shows elegance and craftsmanship even in its abandoned state. (*Photo: Liz Cooke*)

Above right: Interior doorways lead to unexplored spaces in Hoyt House, an abandoned mansion in Staatsburg, New York. (*Photo: Liz Cooke*)

The Dr. Oliver Bronson House is a wonderful example of a unique building being brought back from the brink after being abandoned in the 1970s. Located on the grounds of a penal institution in Hudson, New York, the Bronson House was designated a National Historic Landmark in 2003. Originally built as a residence for Samuel Plumb in 1812, the house and grounds were reimagined by architect Alexander Jackson Davis into a Romantic-Picturesque landscape for Dr. Bronson and his family in 1839 and 1849. The Bronson house is currently undergoing an extensive restoration program by Historic Hudson. With its decaying aesthetic appeal the house has been the set for several movie and catalog shoots.

Located in a picturesque corner of the Catskills and dating from the nineteenth century, Dundas Castle (also called Craig-E-Claire) was a marvel of its time, boasting water, heat, electricity, and an elaborate system of call bells. Used alternately as a private residence and later a summer camp and Masons' lodge, the castle has been unoccupied since the 1920s. It was added to the National Register of Historic Places in 2001. Like many abandoned places, Dundas Castle has a fascinating history. It is said the original owner built the home for his wife who was soon institutionalized and never moved in. The owner left the house untouched, and despite its dilapidated state and graffiti-covered walls, it stands as an enduring monument to one man's profound love for his wife.

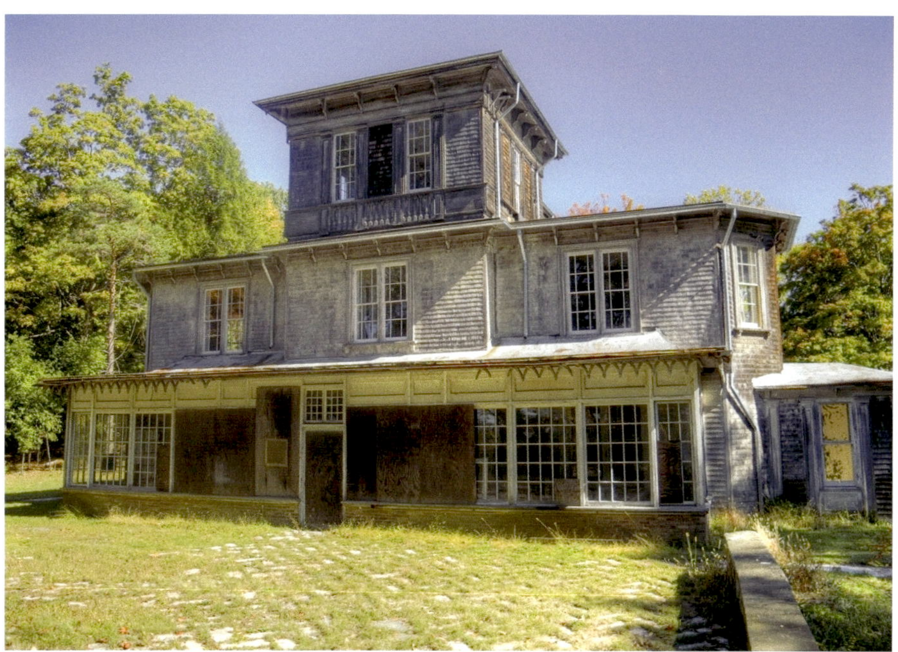

Located in Hudson, New York, on the grounds of a penal institution is the Dr. Oliver Bronson House. Built in 1811–1812, it was abandoned in 1970. Now overseen by Historic Hudson, it was designated a National Historic Landmark in 2003. (*Photo: Andy Milford*)

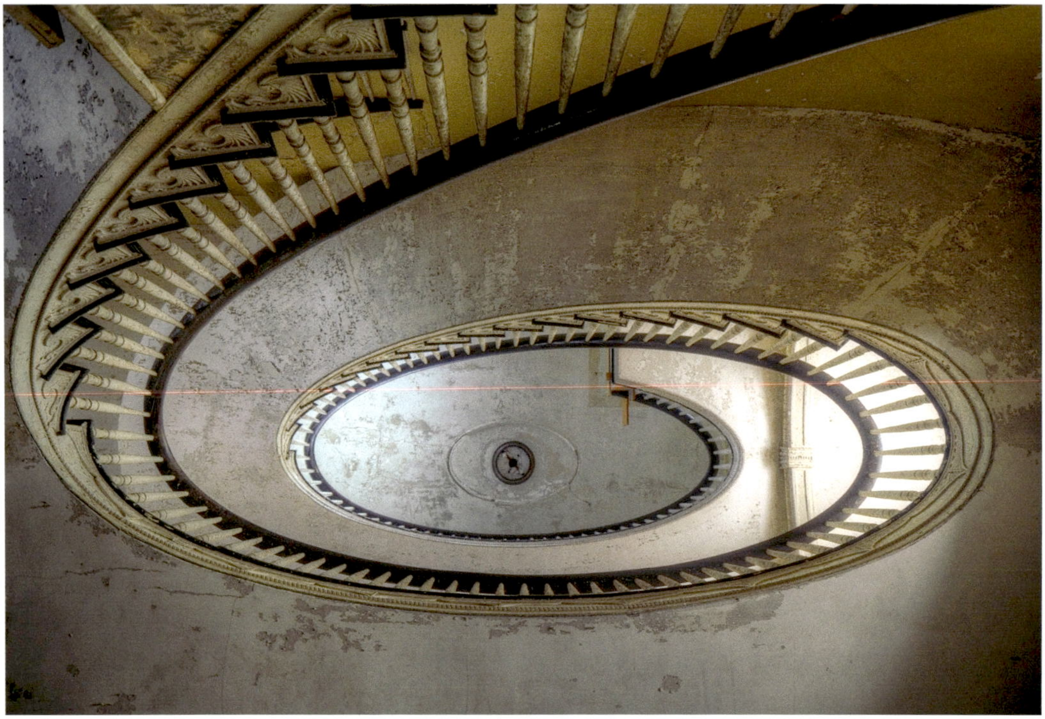

Designed by renowned architect Andrew Jackson Davis, the Bronson house features many exceptional architectural details, including this spiral staircase. (*Photo: Andy Milford*)

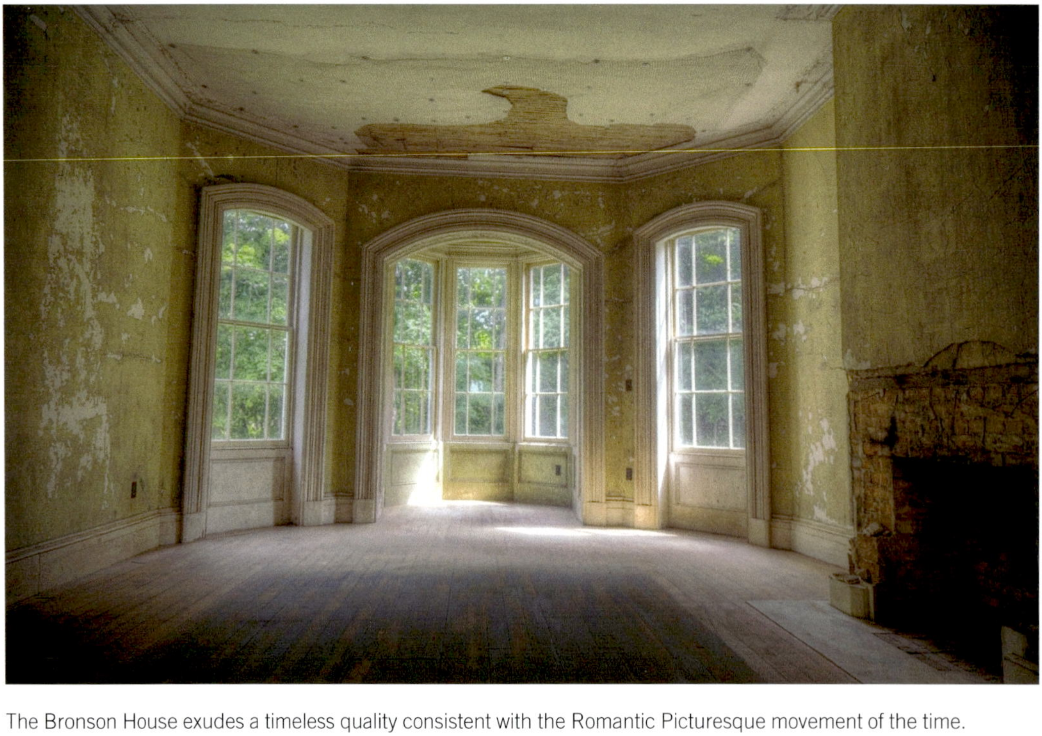

The Bronson House exudes a timeless quality consistent with the Romantic Picturesque movement of the time. (*Photo: Andy Milford*)

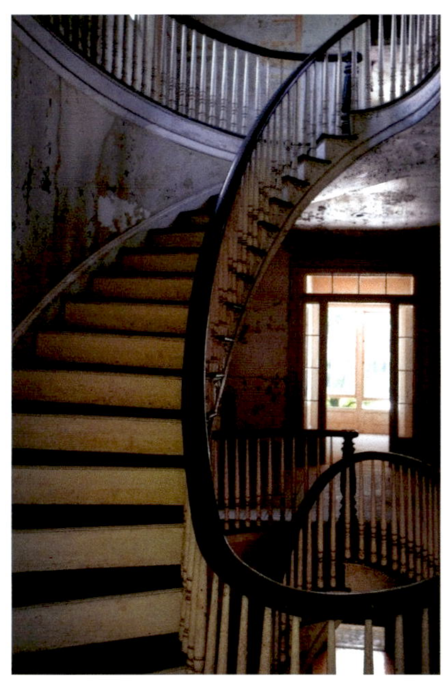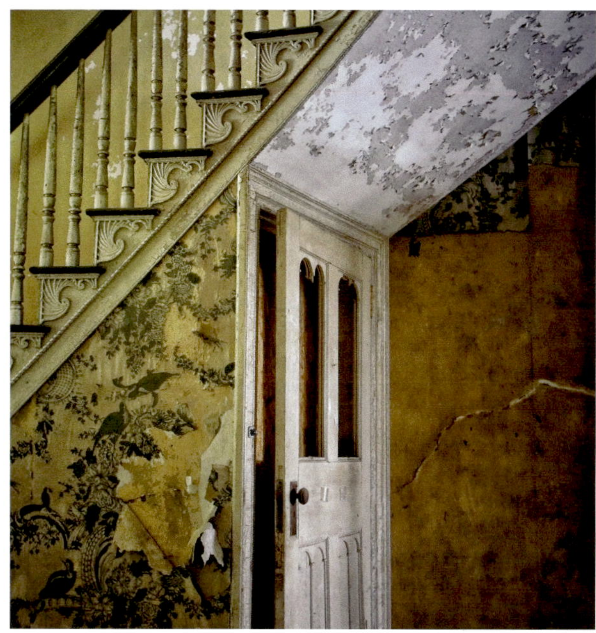

Above left: The Dr. Oliver Bronson House is a three-story Federal-style residence. Despite years of abandonment, many of its original details remain intact, including its impressive spiral staircase. (*Photo: Liz Cooke*)

Above right: Many people are drawn to the Bronson House to experience the aesthetic beauty often found in abandonment. Original architectural details include elaborate wall treatments. (*Photo: Liz Cooke*)

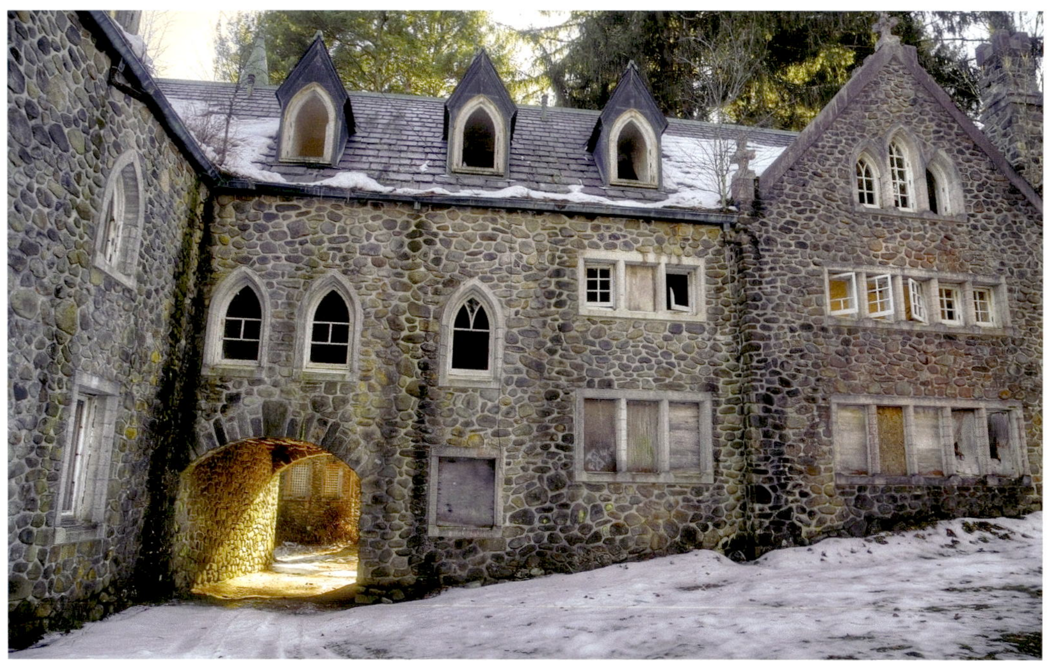

Dundas Castle, also known as Craig-E-Claire, was designed as a private home and later used as a summer camp and Masons' lodge. It has been abandoned for many years and has experienced considerable vandalism. (*Photo: Andy Milford*)

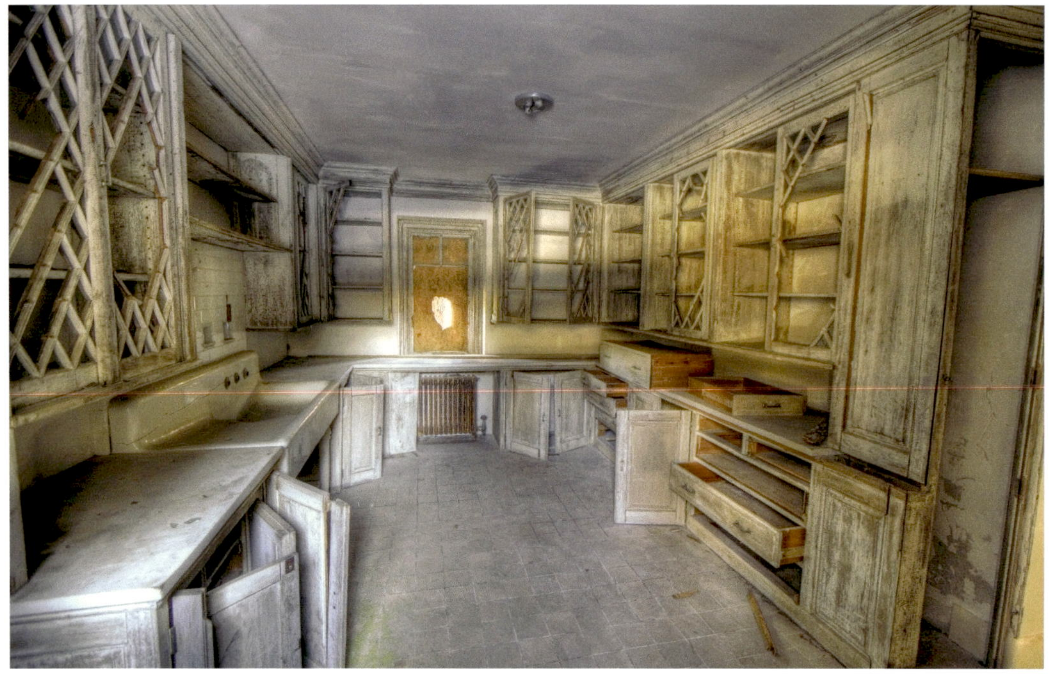

The kitchen at Dundas Castle in Roscoe, New York, was clearly envisioned as a place to be maintained by a staff of servants. (*Photo: Andy Milford*)

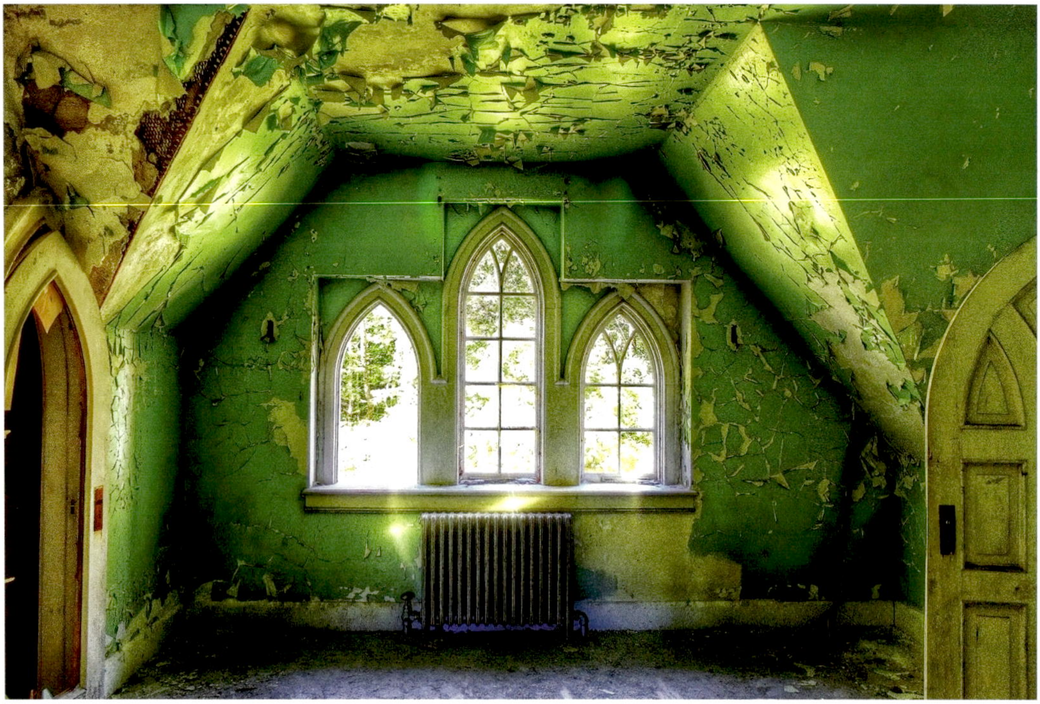

Designed in the Gothic Revival style, Dundas Castle boasts thirty-six rooms, each with its own unique beauty. The property was placed on the National Register of Historic Places in 2001. Its future remains uncertain. (*Photo: Liz Cooke*)

3

SCHOOLS AND RELIGIOUS INSTITUTIONS

As the population of the Hudson Valley grew over the years, so too did the number of schools, churches, and other religious institutions. The need for community gathering places is a universal one, but changing tastes, population trends, and financial demands can cause upheaval. And in this regard, the Hudson Valley is no different from many places where people—and institutions—come and go.

Throughout the Hudson Valley there are any number of abandoned schools, from simple one-room schoolhouses to massive high schools and colleges. Back in their heyday, you might hear students chatting, giggling, and planning their futures. You might hear teachers encouraging good study habits or sternly warning against misbehavior. Everything would be as it is supposed to be—except these everyday occurrences will not last. For unknown reasons, many schools in the Hudson Valley opened, flourished, and then closed with little to no explanation or ceremony.

Quite possibly the Hudson Valley's most notable (and iconic) abandoned building was located in Millbrook, New York. Until its final days in 2021, Halcyon Hall, the grand centerpiece of Bennett College, stood as a hauntingly beautiful shell of its former self, attracting sightseers, screenwriters, photographers and a large flock of vultures that found eerily perfect perches.

For many years, this imposing structure, once a thriving women's college, lay abandoned and forgotten. Coming upon it, you were immediately struck by its massive stone facade and crumbling wooden beams. It gave the impression of something out of a horror movie.

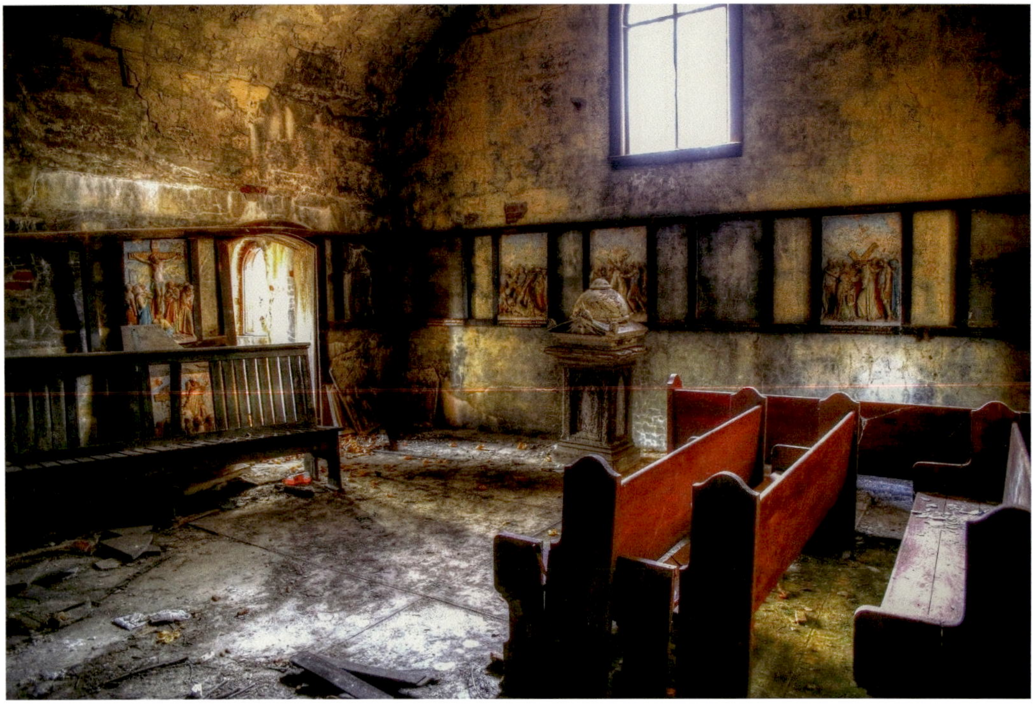

Situated in a Poughkeepsie, New York, cemetery, this abandoned chapel was once a sacred place. When a modern chapel was built on the grounds, this beautiful structure was left in a state of neglect. (*Photo: Liz Cooke*)

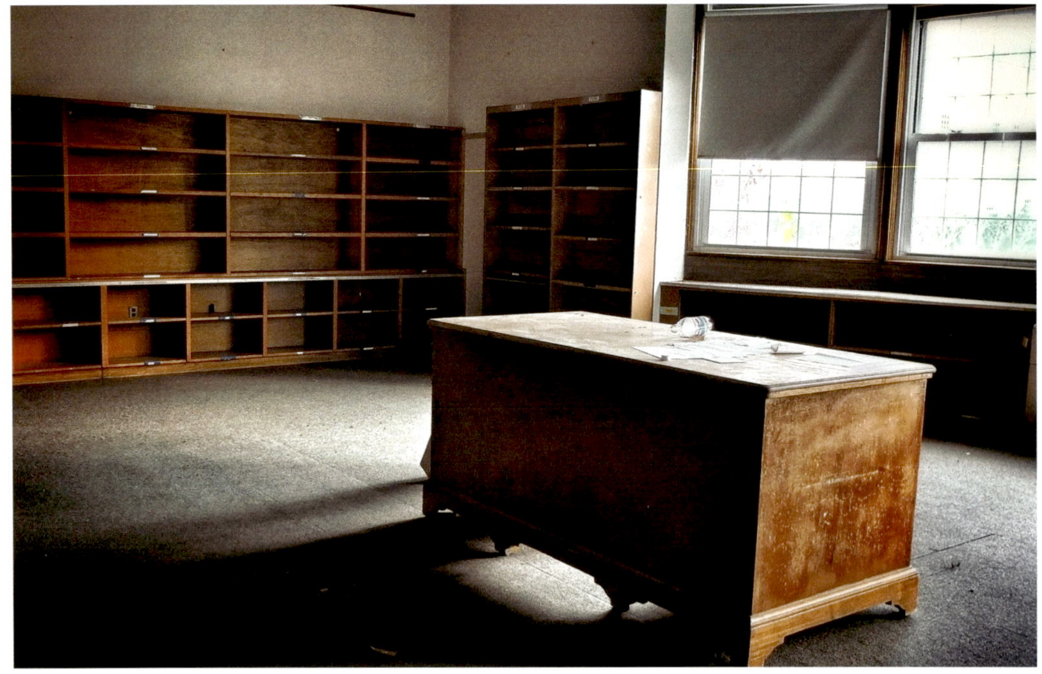

Built in 1932 and vacant since 1999, the Roeliff-Jansen School in Copake, New York, was the alma mater of hundreds of local children. Plans are under consideration to convert the almost 100,000 square foot property into affordable housing. (*Photo: Liz Cooke*)

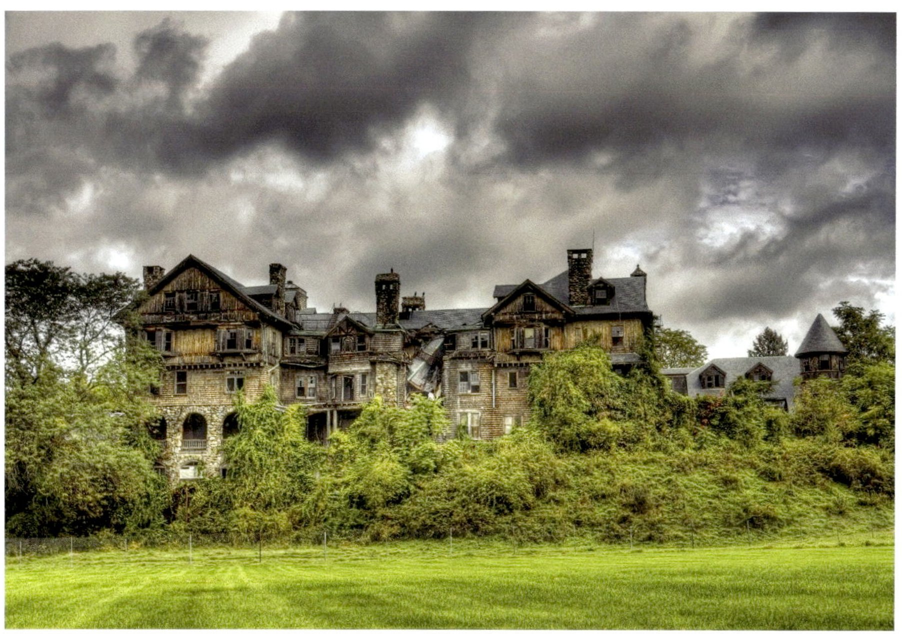

The embodiment of abandonment in the Hudson Valley was Halcyon Hall, the largest building on the Bennett College campus in Millbrook, New York. After years of neglect and community debate, Halcyon Hall was demolished in 2021. (*Photo: Andy Milford*)

Founded by May Bennett in 1890, Bennett College (earlier known as Bennett School for Girls) moved to a 22-acre campus in Millbrook in 1907 and operated there through 1978. As coeducation came into vogue, the fortunes of Bennett College fell. To keep up with the times, the Kettering Science Center was built in 1972, but rather than improving the school's finances, it caused further financial burdens that resulted in the school's bankruptcy. Halcyon Hall fell victim to the forces of time and nature. With no maintenance, it was frequently flooded, hastening its demise. For many years, and much to the dismay of the affluent residents of Millbrook, the grand structure stood vacant, deteriorating for all to see. The building was slated for demolition in 2011 and 2012 and finally was demolished in 2021.

The tranquil, rural setting of Dutchess County provided an ideal backdrop for religious life. Throughout the Hudson Valley there are any number of abandoned and once-abandoned religious institutions including the former home of the Community of Sisters of St. Mary convent in Peekskill, New York. These once-vibrant structures—now in various states of decline and decay—speak to a time when churchgoing was an important part of everyday life. The decline and eventual abandonment of many religious institutions in Dutchess County can be attributed to several factors, including demographic changes, economic pressures, and shifts in religious practice. One example of a church whose time has come and gone is the Shookville Church, once

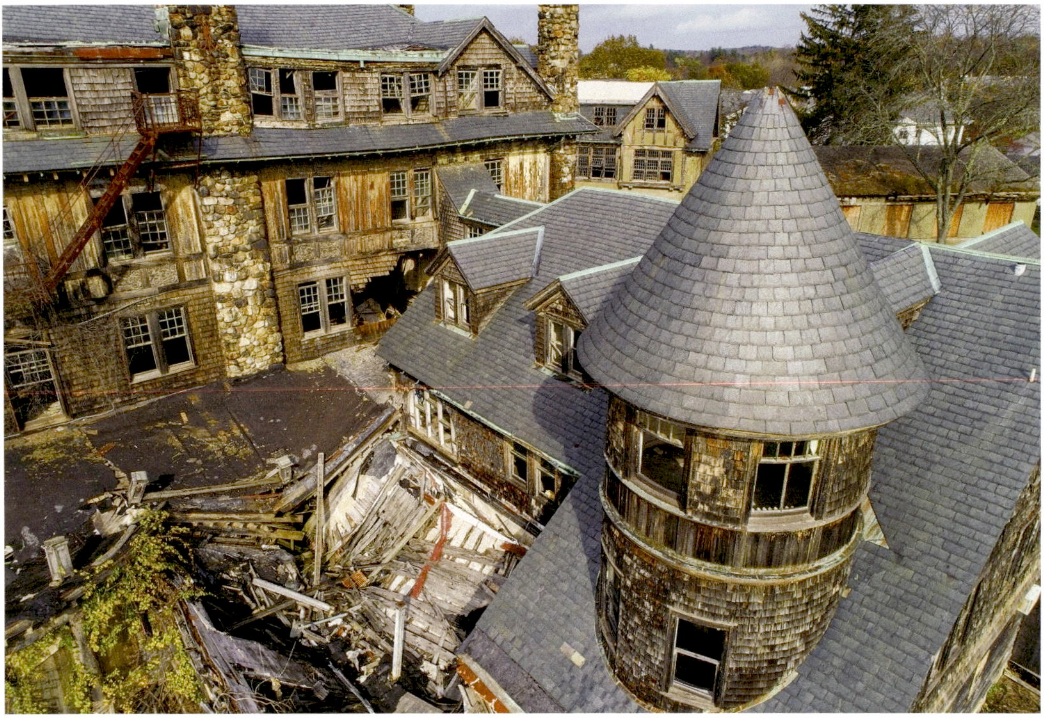

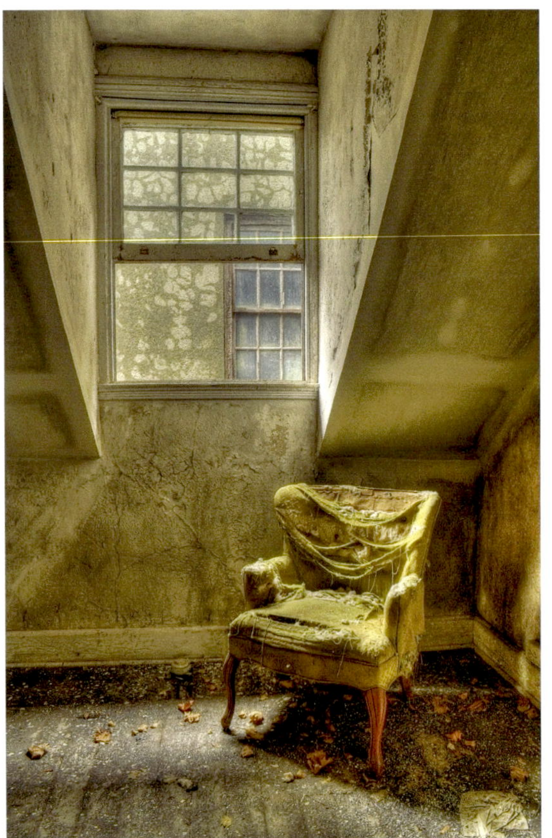

Above: An aerial view of Halcyon Hall, built in 1870, abandoned in 1978 and demolished in 2021. (*Photo: Andy Milford*)

Left: The iconic chair in Halcyon Hall, the grand centerpiece of Bennett College, Millbrook, New York. (*Photo: Andy Milford*)

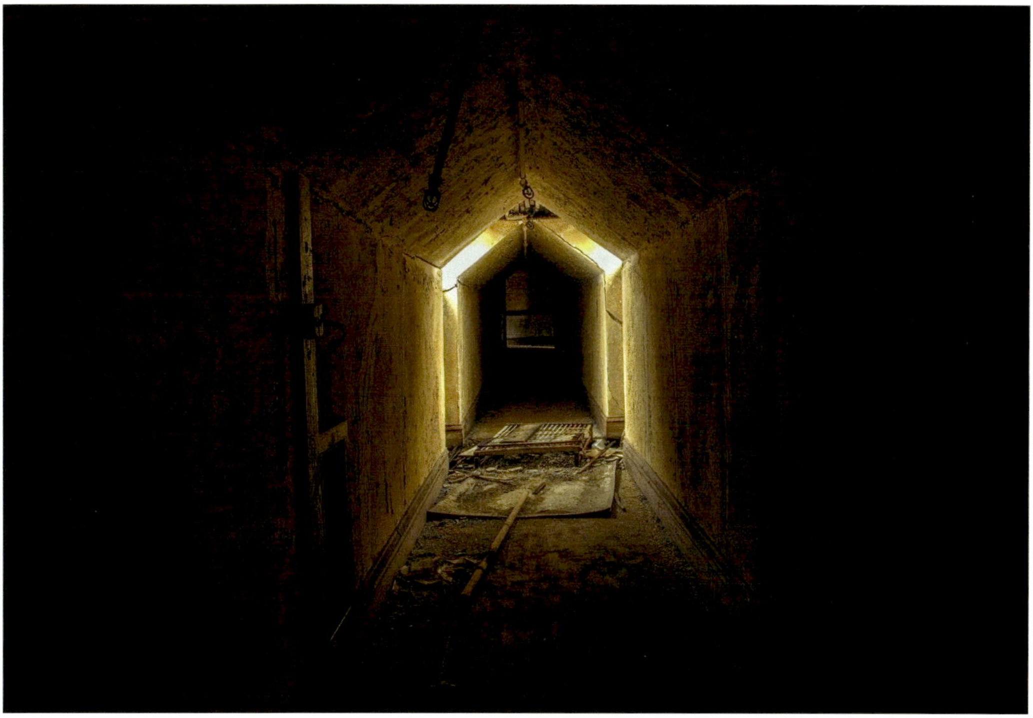

Shadow and light in the hallway of Halcyon Hall, a "finishing school" where daughters of the elite took lessons in art, history, dance, and domestic arts. (*Photo: Andy Milford*)

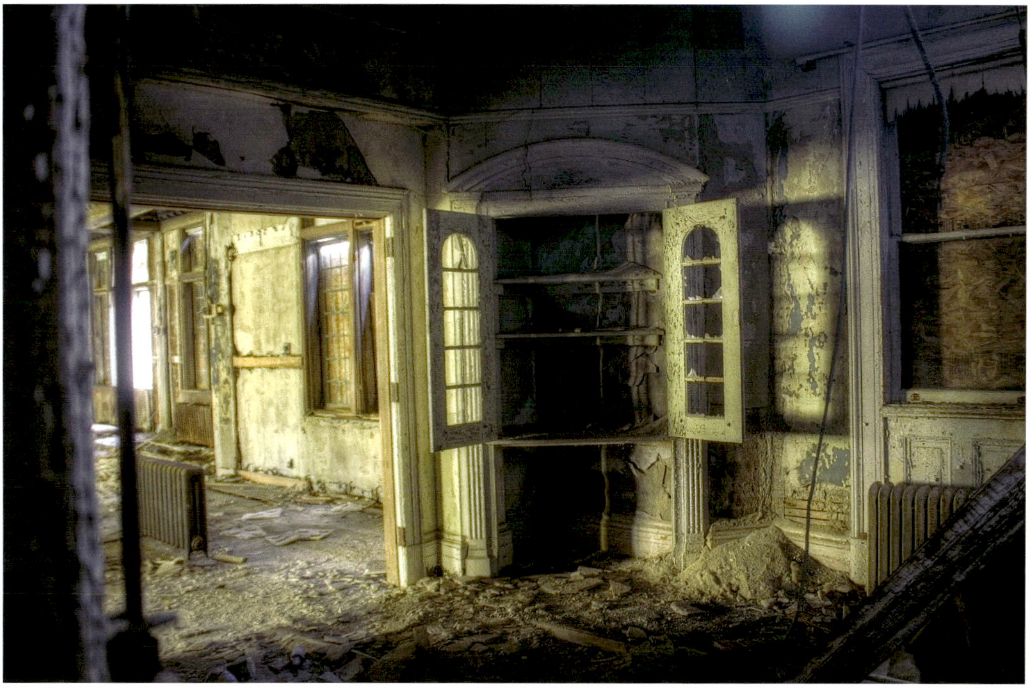

Crafted with the finest materials of the time, Halcyon Hall was once bustling with activity. Former students still recall with great fondness time spent on this once-grand campus. (*Photo: Andy Milford*)

a center of community life in Dutchess County and now a mere ruin. In Poughkeepsie, New York, nestled in a large cemetery is an abandoned chapel with exquisite wall sculptures depicting the stations of the cross, and with pews arranged as if mourners might come any moment. Many other abandoned churches can be found throughout the Hudson Valley, although still others have been reborn as private residences, arts and entertainment venues, hotels, and retail establishments.

While these places of worship have their own stories to tell, there are others that have a more profound impact. Among these is the Dutch Reformed Church in Newburgh, New York. Established in 1835, it was a prominent place of worship for the Dutch Reformed community. Like many historic structures, the Dutch Reformed Church faced a decline in the latter half of the twentieth century and was threatened with demolition as part of Newburgh's urban renewal efforts. It was placed on the National Register of Historic Places in 1970, sparing it from the wrecking ball, and in 2001 was named a National Historic Landmark. Despite strenuous efforts by preservationists to and small gestures toward repair, the structure continues to suffer from neglect and exposure to the elements. An adaptive reuse plan is under way and could bring new life back to this worthy structure.

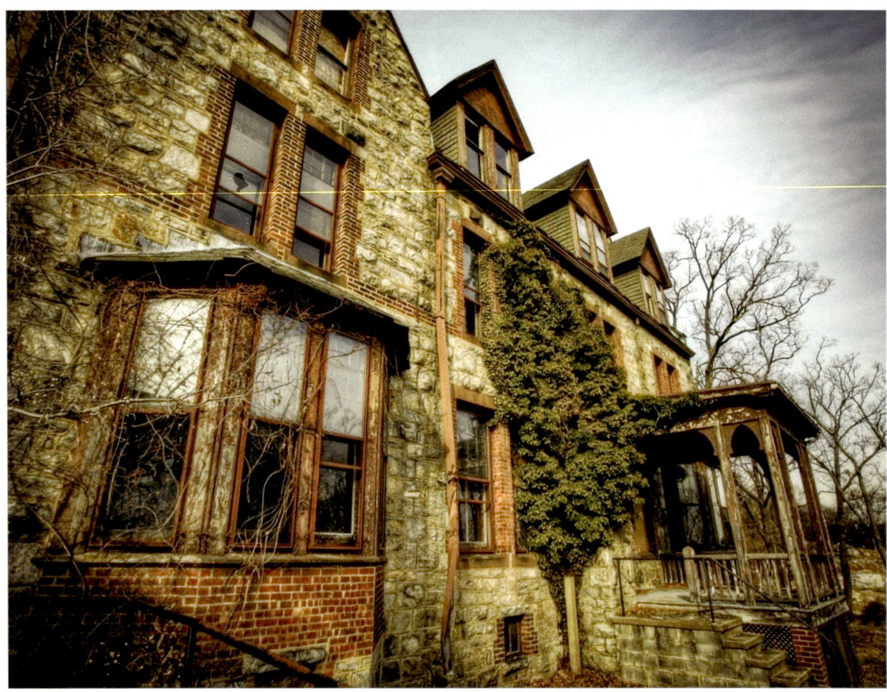

Located in the historic Fort Hill section of Peekskill, New York, the former Community of St. Mary convent was vacant for many years. It has since been reborn as The Abbey Inn and Spa, a luxury retreat. (*Photo: Liz Cooke*)

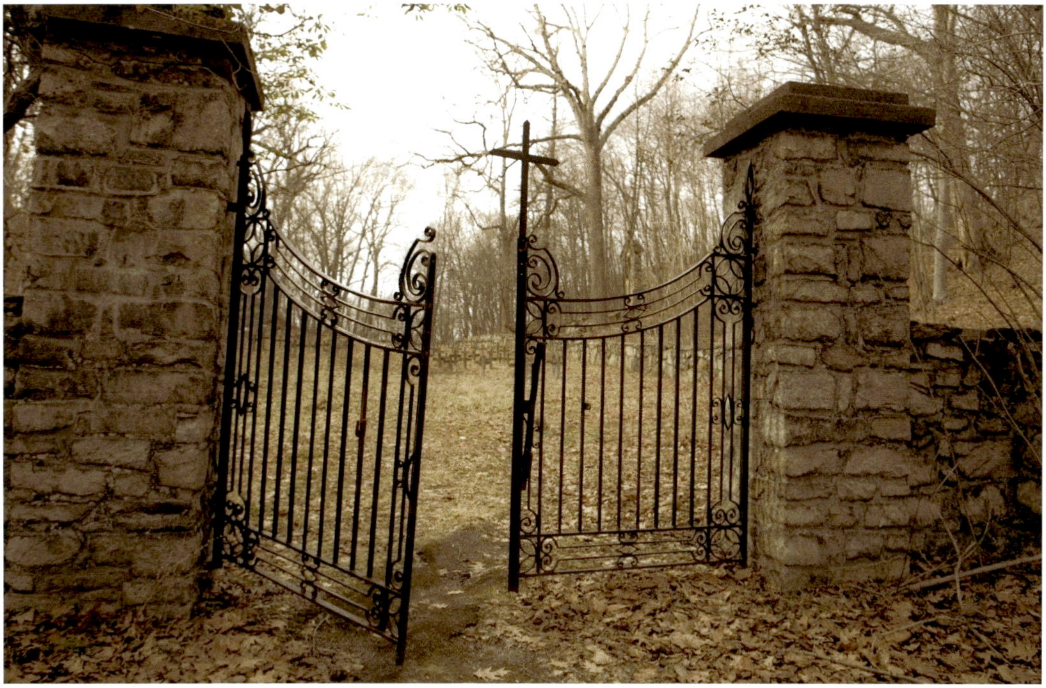

This cemetery was once the final resting place for many of the sisters who called the Community of St. Mary convent home. (*Photo: Liz Cooke*)

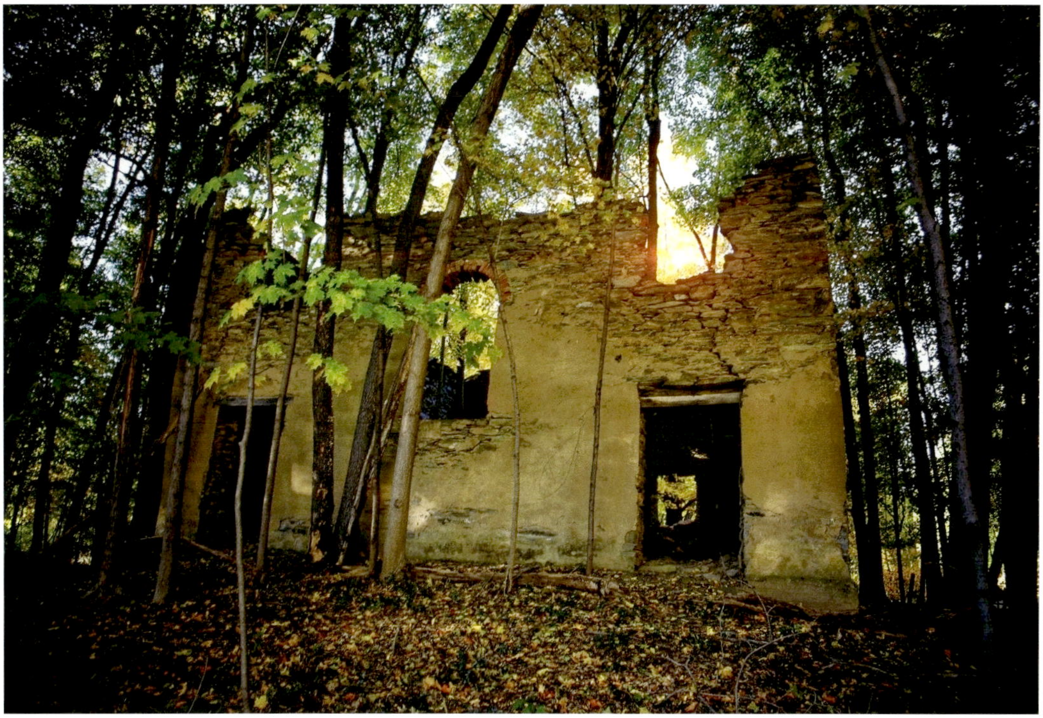

This crumbling ruin is all that remains of the Shookville Church, once the spiritual center of the tiny hamlet of Shookville near Milan, New York. (*Photo: Liz Cooke*)

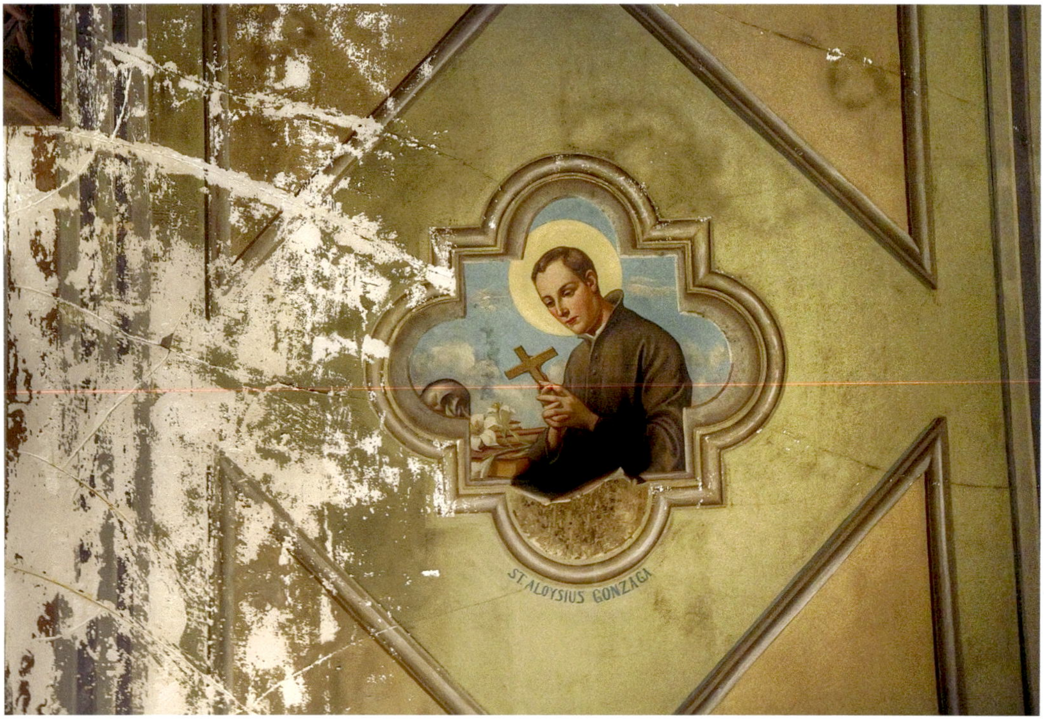

Religious imagery covers every wall in this abandoned Catholic school. (*Photo: Andy Milford*)

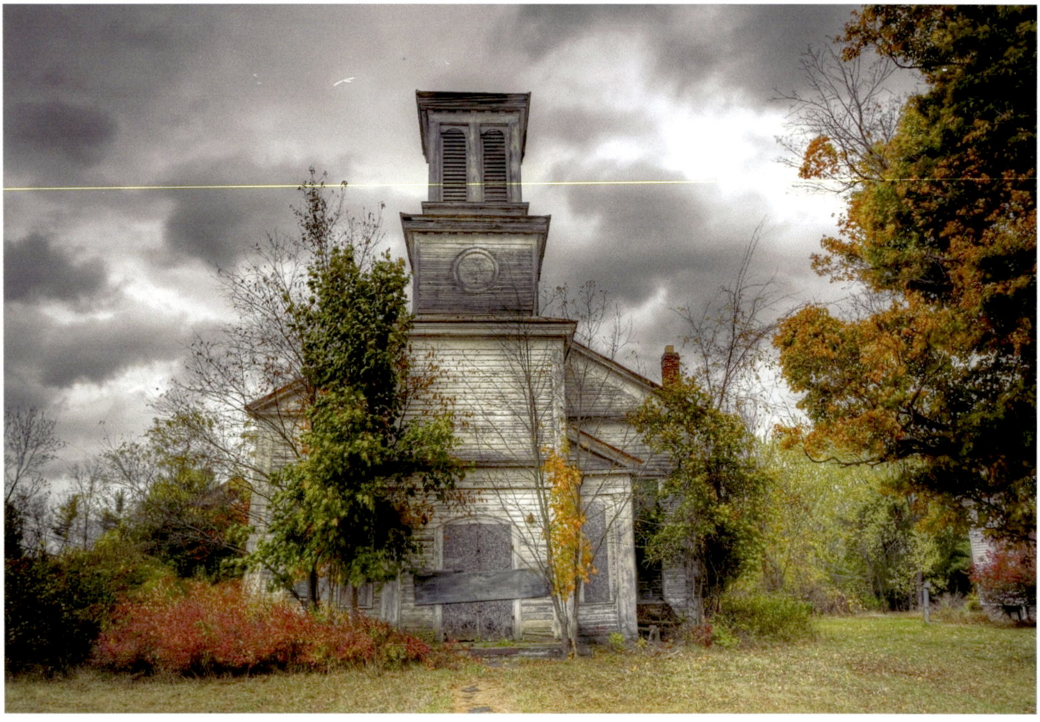

The Hudson Valley has any number of abandoned churches and other places of worship, each with a different story to tell of weddings, prayer services, baptisms, and other religious activities. (*Photo: Liz Cooke*)

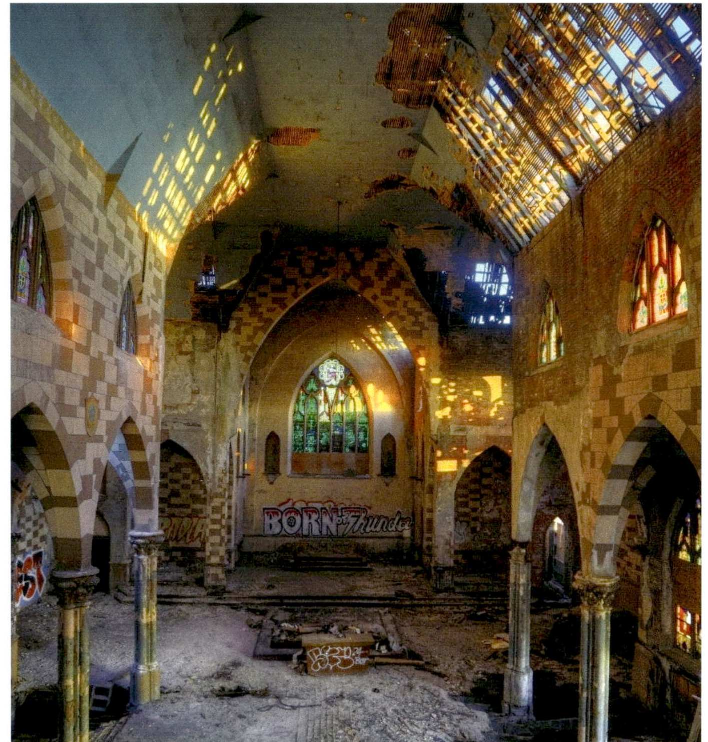

Right: Stained glass casts a colorful glow in this abandoned Catholic church. (*Photo: Andy Milford*)

Below: Located in Newburgh, New York, the Dutch Reformed Church played a prominent role in the spiritual life of residents before facing demolition in the 1960s. It was added to the National Register of Historic Places in 1970. (*Photo: Liz Cooke*)

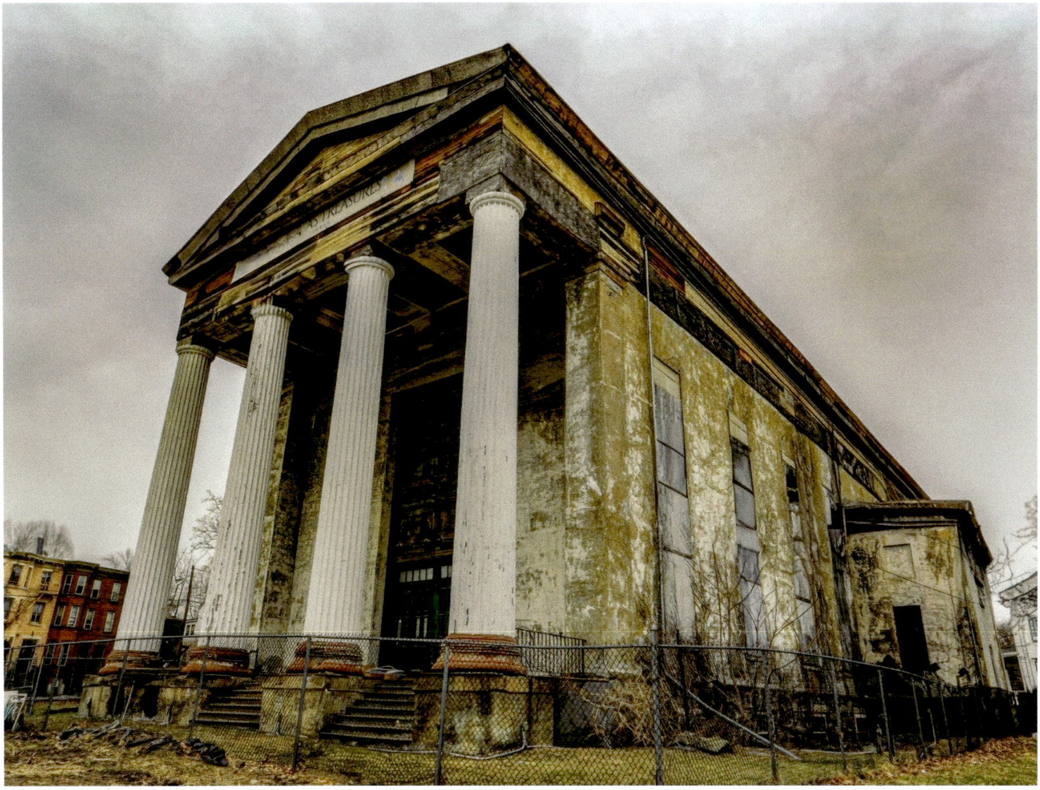

4

HOTELS, MOTELS, AND "BORSCHT BELT" RESORTS

With its picturesque landscapes and proximity to urban centers, the Hudson Valley has long been a prime destination for vacationers. From modest boarding houses and cozy bungalows to grand hotels and even the famous resorts of the "Borscht Belt," the Hudson Valley truly offered something for everyone. But over time, even the grandest of hotels and lowliest of motels fell into disuse and disrepair, and what remains are ghostly reminders of what once was.

The term "Borscht Belt" refers to the area in the Catskill Mountains where numerous resorts catered primarily to Jewish vacationers from the 1920s to the 1970s. Offering kosher cuisine and other amenities, the popularity of Borscht Belt hotels peaked in the 1950s and 1960s.

During its heyday, the Borscht Belt was a cultural hub, fostering the development of Jewish-American entertainment and influencing American comedy. Comedians such as Jerry Lewis, Milton Berle, and Mel Brooks honed their craft in the resorts' nightclubs, often performing multiple shows in a single night. Musicians, including Duke Ellington and Louis Armstrong, also performed in the Catskills, contributing to the vibrant nightlife that was a hallmark of the Borscht Belt experience, even inspiring the movie *Dirty Dancing*.

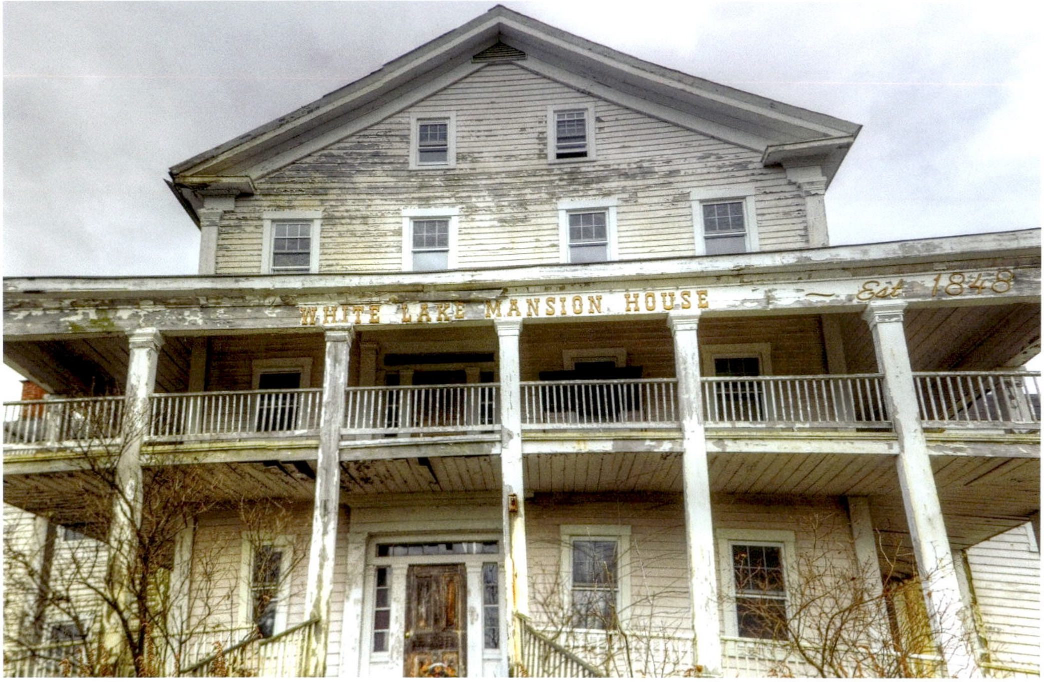

Built in 1848, the White Lake Mansion Hotel in the Catskills town of White Lake, New York, was once a destination of choice for visitors from New York City. It was abandoned in 2000. (*Photo: Andy Milford*)

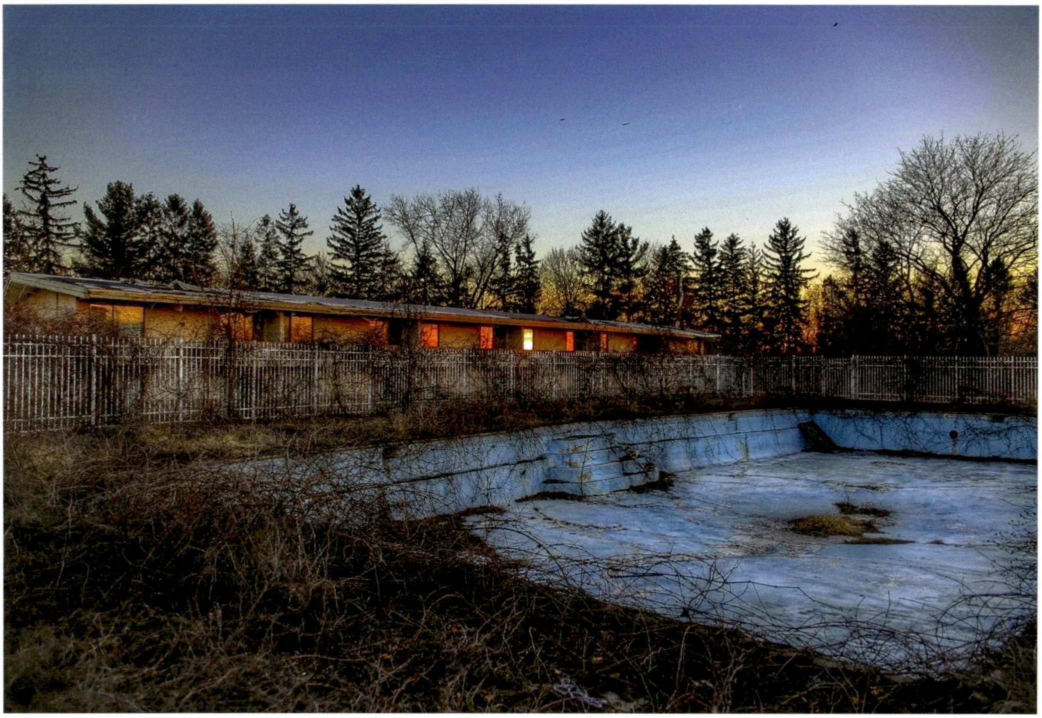

Visitors to the Hudson Valley and other upstate regions often take the busy New York Thruway. This abandoned motel is one of many that can be found nearby. (*Photo: Liz Cooke*)

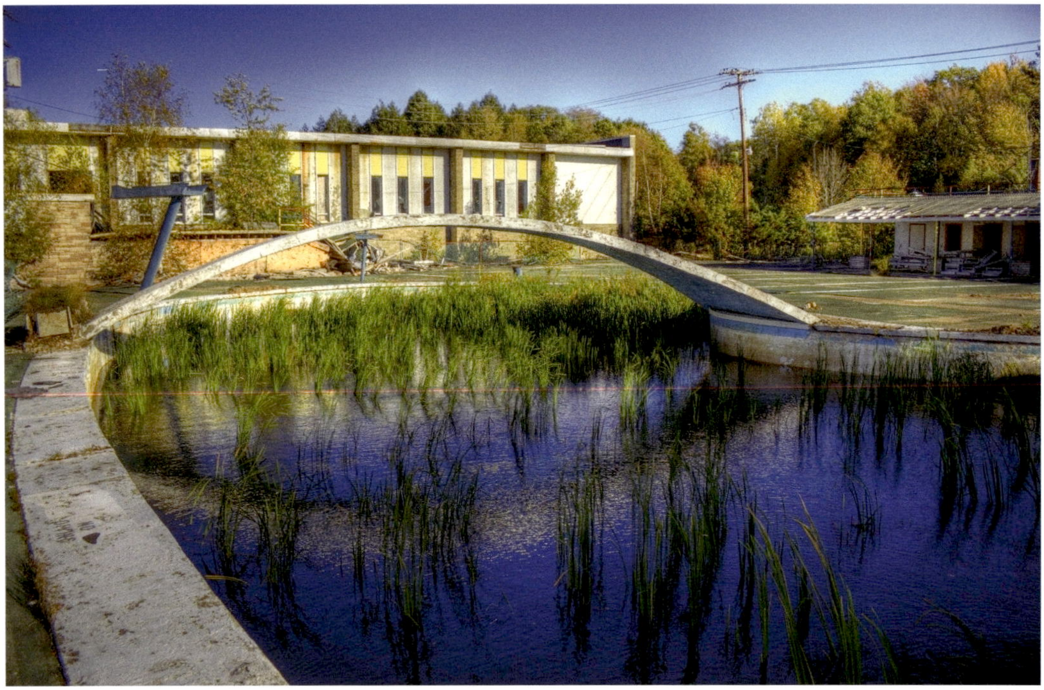

Established in 1933 in South Fallsburg, The Pines catered predominantly to Jewish families from New York City. Its unique design included a bridge over a pool, among many other "mid-century" features. (*Photo: Andy Milford*)

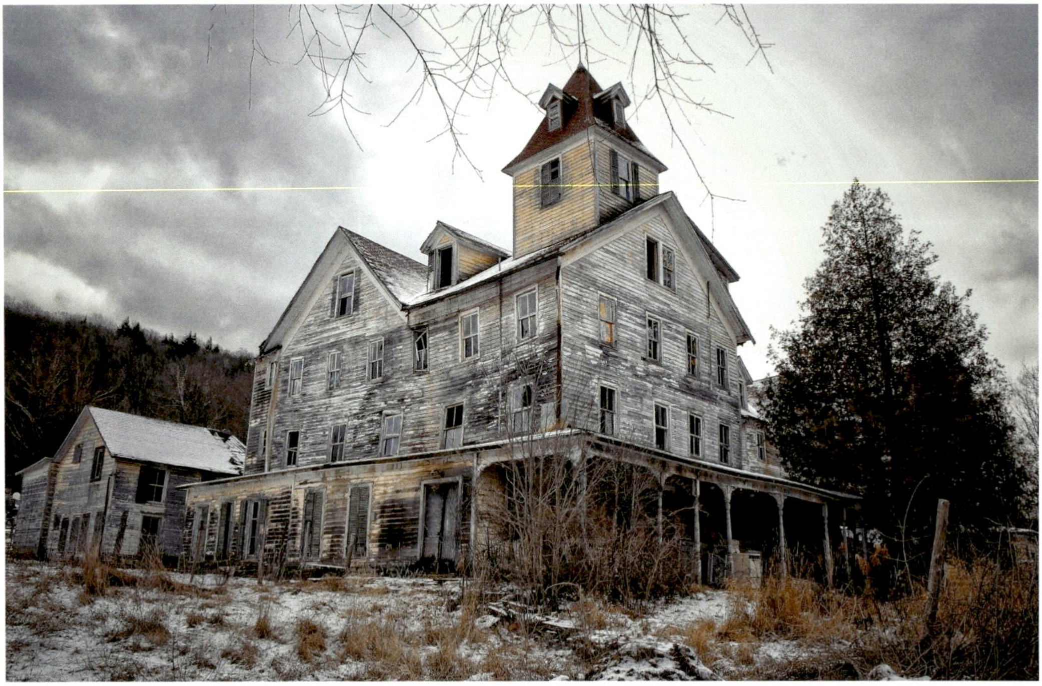

The Cold Spring Hotel in Tannersville, New York, was built in the 1890s. According to a 1904 advertisement, the property offered guests such desirable amenities as indoor plumbing and fire extinguishers. The hotel was abandoned in the 1980s. (*Photo: Liz Cooke*)

Grossinger's Catskill Resort Hotel, perhaps the most famous of the Borscht Belt resorts, began as a modest farmhouse that Jennie Grossinger and her family turned into a boarding house in 1914. Over the decades, Grossinger's expanded into a luxurious resort covering over 1,200 acres with more than thirty-five buildings. By the mid-twentieth century, it featured amenities such as a golf course, an Olympic-sized swimming pool, and a ski slope. Grossinger's became a symbol of the Borscht Belt's prosperity, drawing thousands of visitors each year until its closure in 1986.

The Homowack Lodge started as a small hotel in the early 1920s, eventually growing into a popular resort. Known for its friendly atmosphere and excellent food, the Homowack attracted families who sought a relaxing and entertaining summer vacation. The resort offered a range of activities, including bowling, swimming, hiking, and tennis, along with evening entertainment. Despite its popularity, the Homowack Lodge could not sustain its success and closed in the early 2000s.

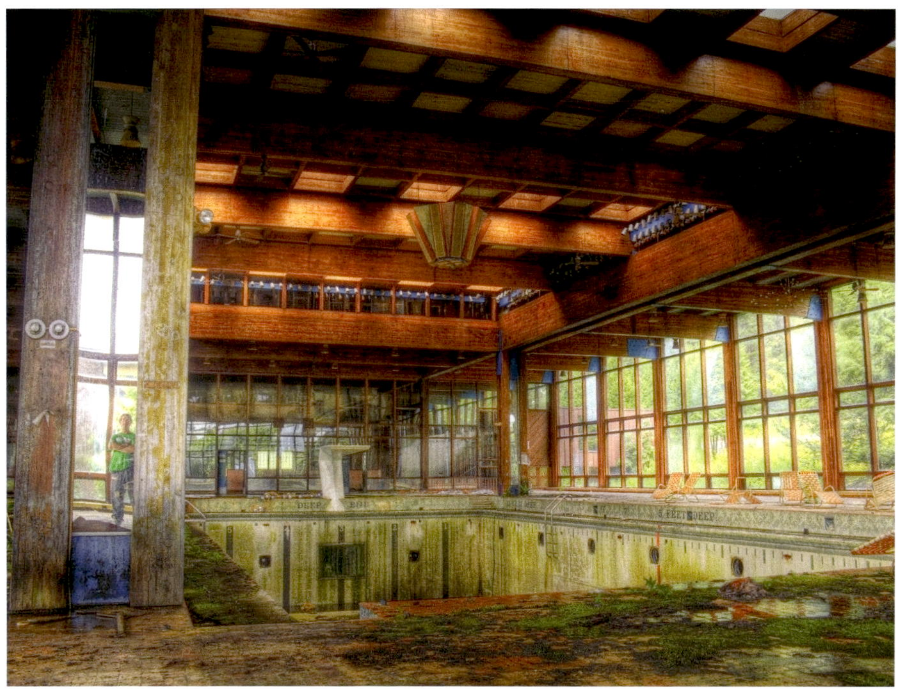

Grossinger's Resort, one of the largest Borscht Belt resorts, offered kosher cuisine to primarily Jewish guests, as well as a range of luxurious amenities including this expansive indoor pool. (*Photo: Andy Milford*)

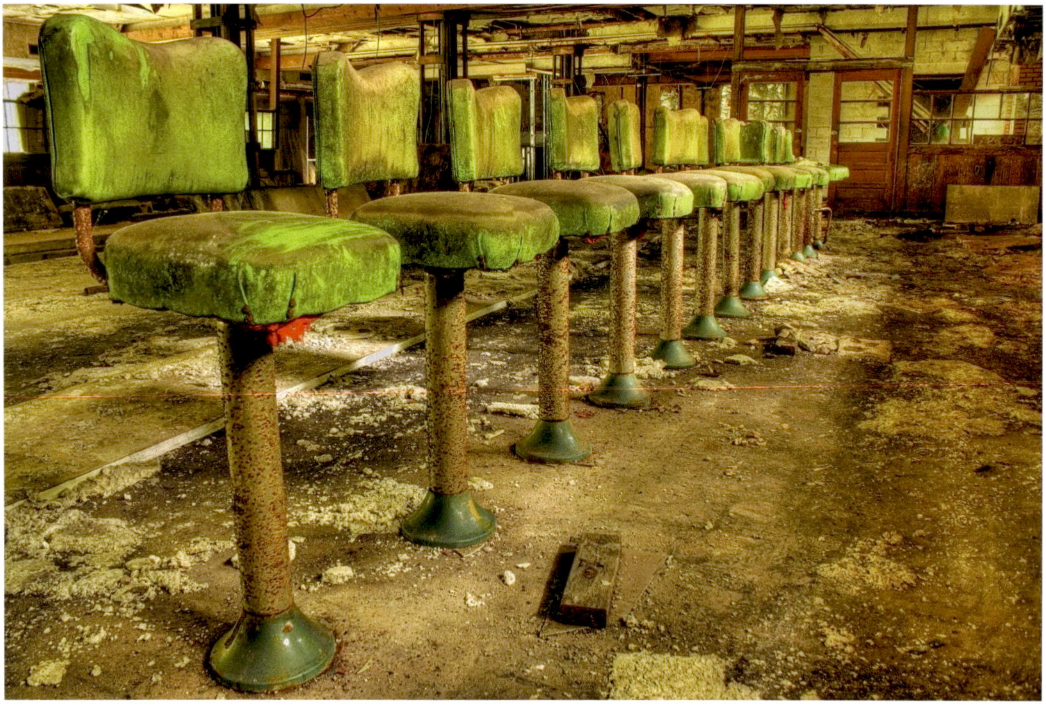

The enormous dining room at Grossinger's Resort in Liberty, New York, was able to seat 1,300 guests. These stools were all that remained of one of its more casual dining places. (*Photo: Andy Milford*)

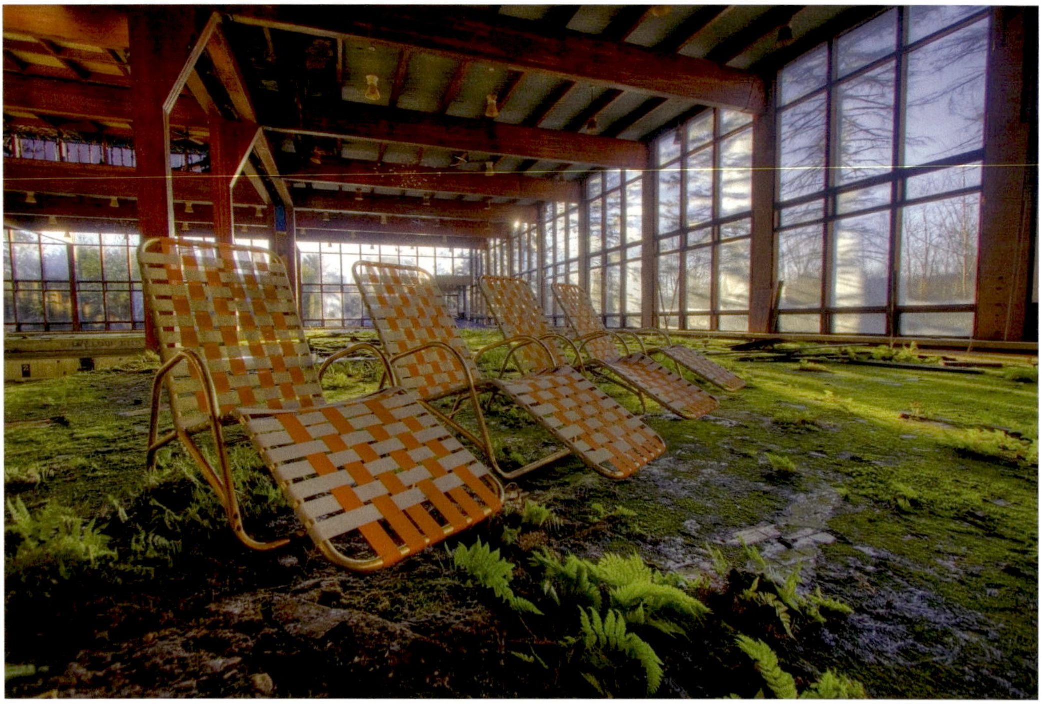

These chaise lounges invited guests to relax and linger poolside at Grossinger's Resort in Liberty, New York. (*Photo: Andy Milford*)

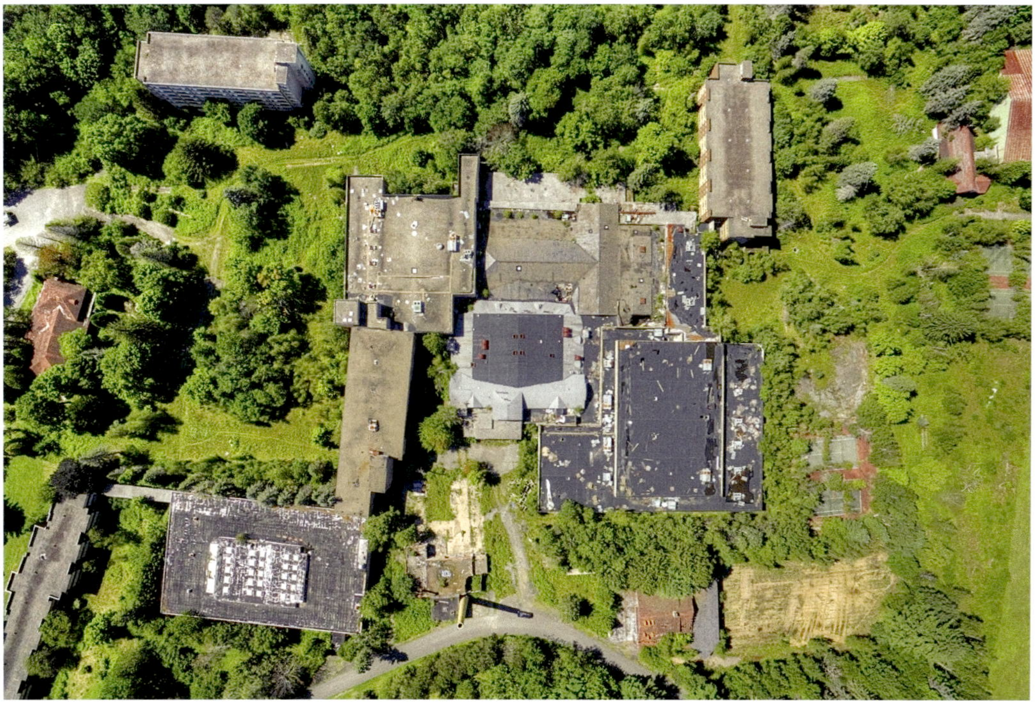

An aerial view of the enormous Grossinger's Resort property, which at its peak had thirty-five buildings. Most of these buildings were demolished by 2018. (*Photo: Andy Milford*)

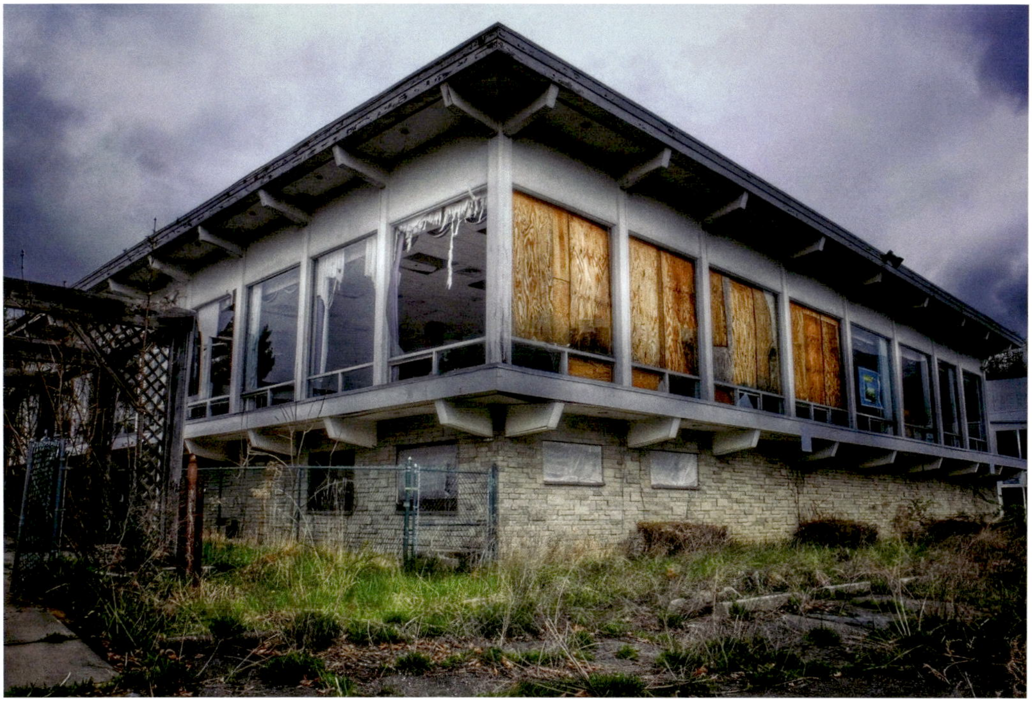

Exterior shot of the abandoned Homowack Lodge in Spring Glen, New York. Formerly a family resort, it later became a Hasidic school for girls before its closure in 2009. (*Photo: Liz Cooke*)

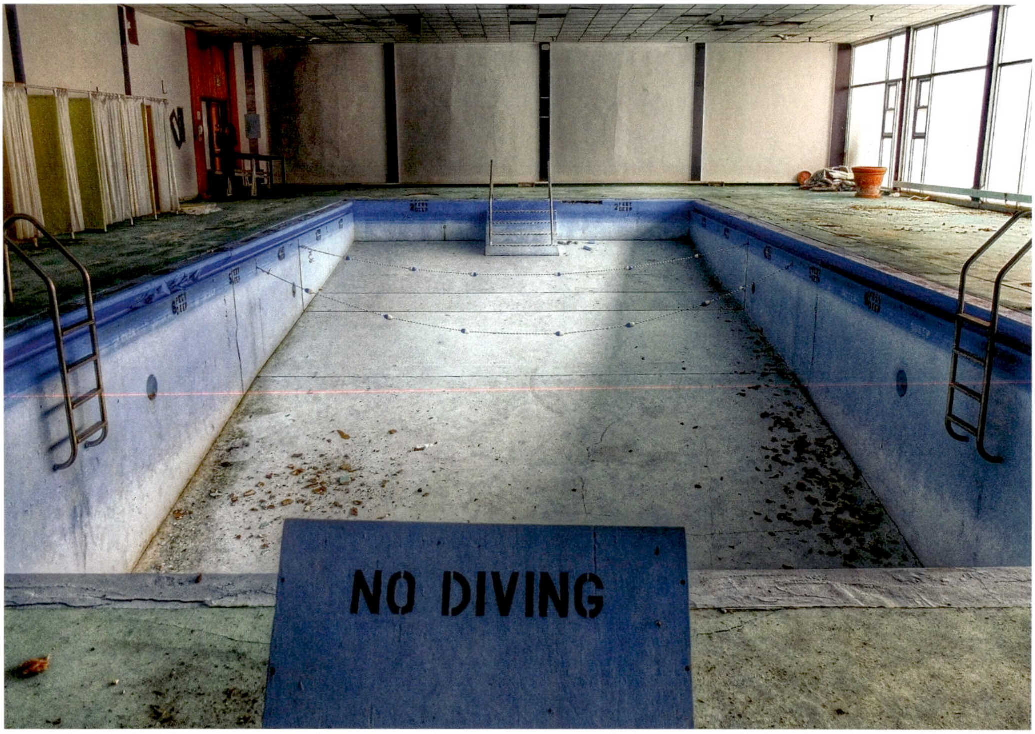

One of several pools at the Homowack Lodge in Spring Glen, New York. (*Photo: Liz Cooke*)

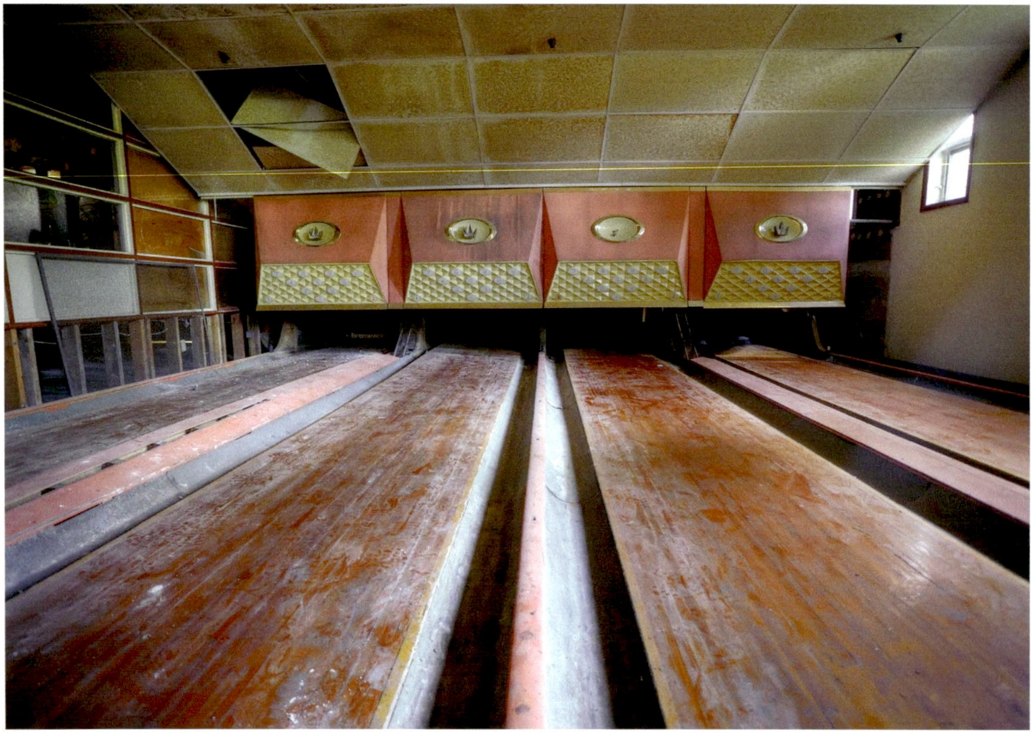

A fully functioning bowling alley was one of many amusements at the Homowack Lodge. (*Photo: Andy Milford*)

The Pines Hotel, once a vibrant jewel in the heart of New York's Borscht Belt, epitomized the mid-twentieth-century allure of the Catskills as a premier vacation destination. Established in 1933 in South Fallsburg, New York. The Pines catered predominantly to Jewish families from New York City. In the 1950s and 1960s, the hotel featured amenities such as an Olympic-sized swimming pool, a golf course, and a bustling nightclub where performers like Buddy Hackett and Tony Bennett entertained guests. However, the late twentieth century saw a decrease in patronage and The Pines Hotel closed its doors in 1998.

Inspired by Native American culture, the Tamarack Lodge had a humble beginning as a six-room boarding house in 1903, expanding significantly by the 1920s and reaching its peak of popularity in the 1950s and 1960s. With diverse entertainment including The Who and Janis Joplin, the Tamarack was the place to be for a good time in the Catskills. The Tamarack Lodge offered various recreational activities and events, such as dances, theater performances, and sports. Like many of its counterparts, the Tamarack Lodge struggled to adapt to changing vacation trends and eventually closed in 2000. In 2012, a fire destroyed most of the site.

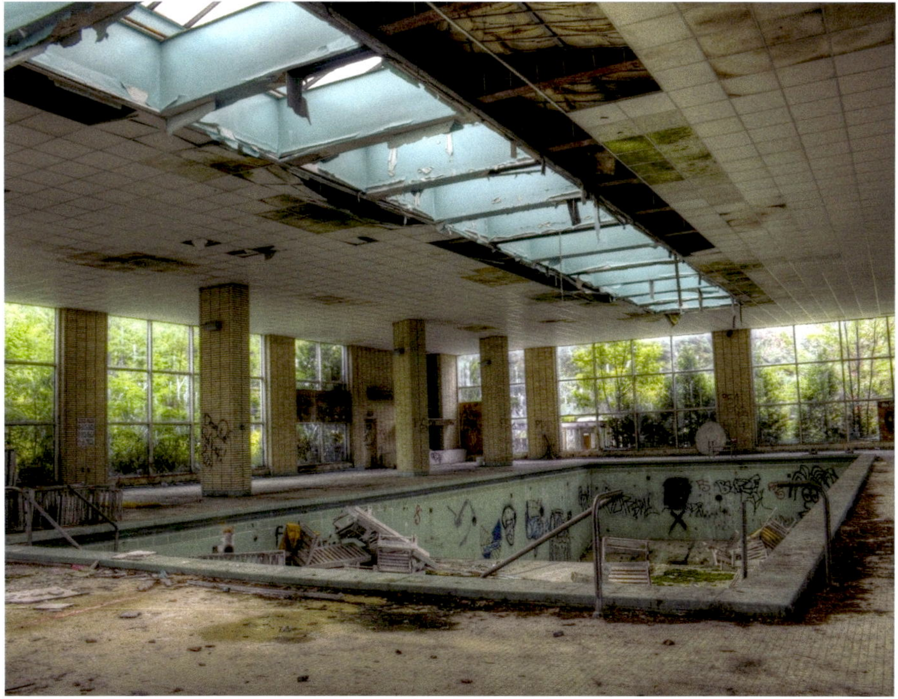

Guests of The Pines resort in South Fallsburg, New York, enjoyed indoor and outdoor pools and even an ice-skating rink. (*Photo: Andy Milford*)

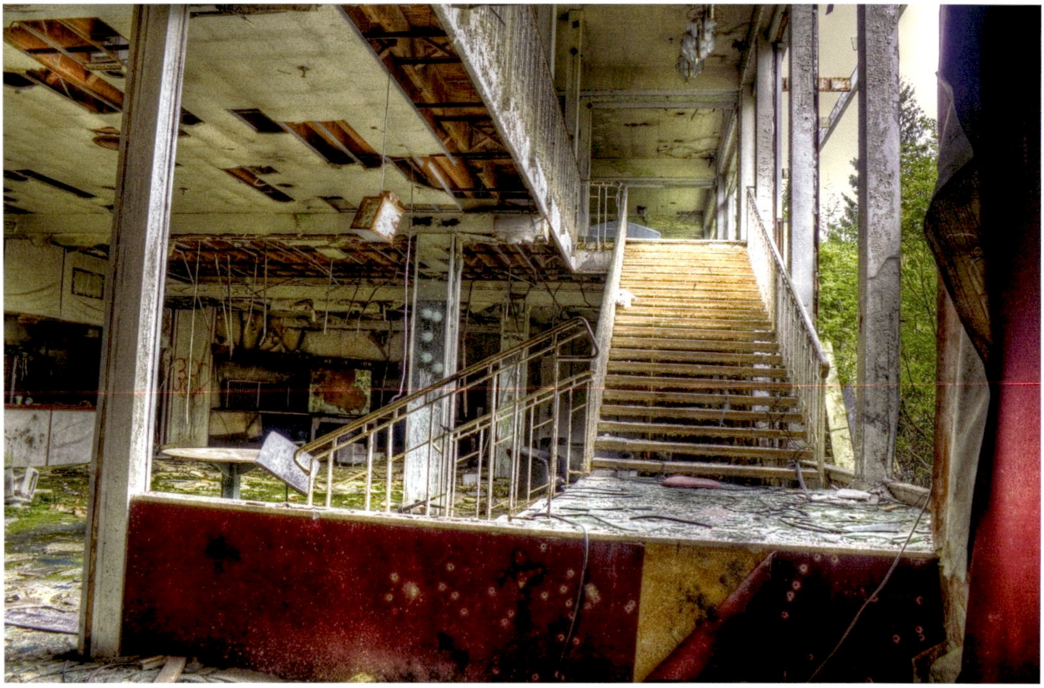

Since closing in 1998, most of the buildings on The Pines' 96-acre site have fallen into extreme disrepair. (*Photo: Andy Milford*)

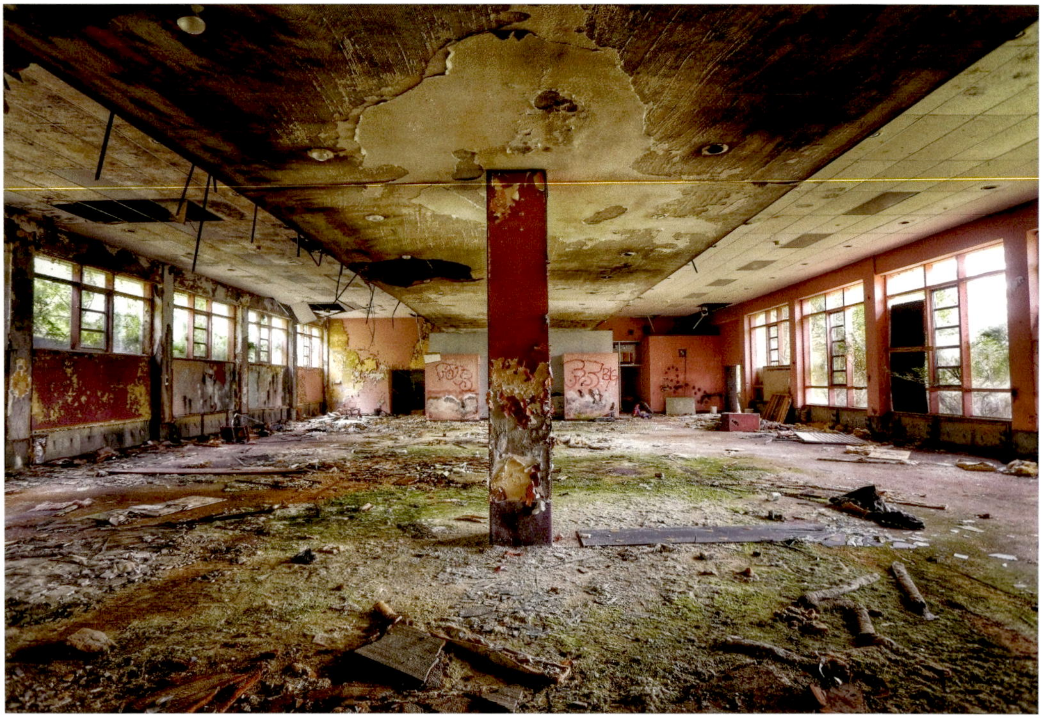

The abandoned Pines resort stood for many years before a 2023 fire destroyed many of the remaining structures. (*Photo: Liz Cooke*)

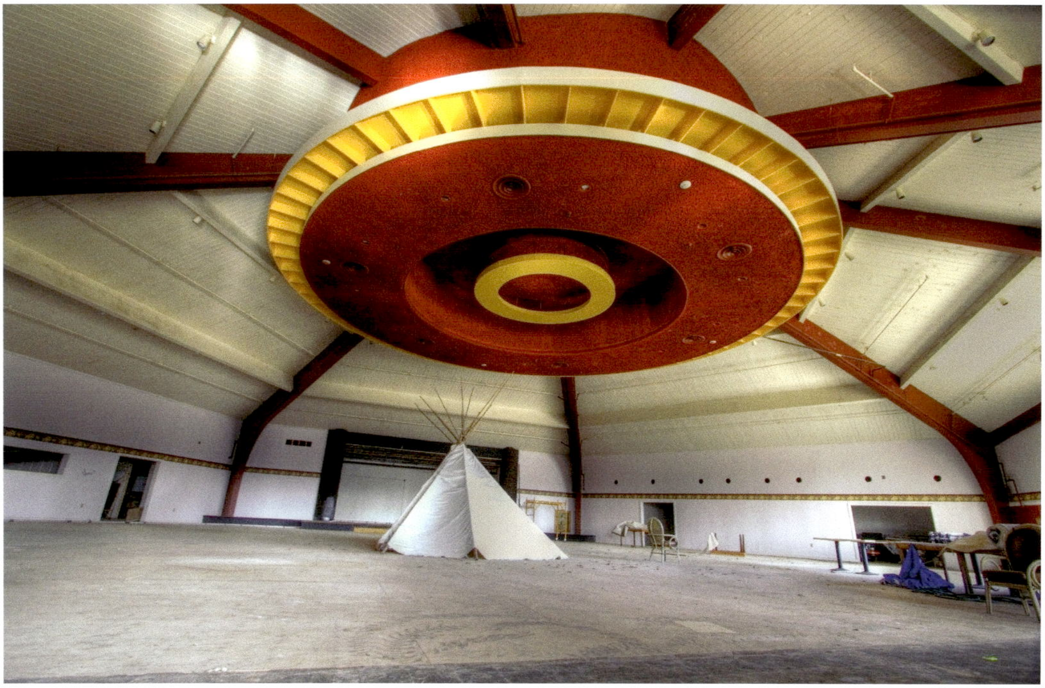

The Tamarack Lodge in Greenfield Park, New York, borrowed many themes from Native American culture. (*Photo: Andy Milford*)

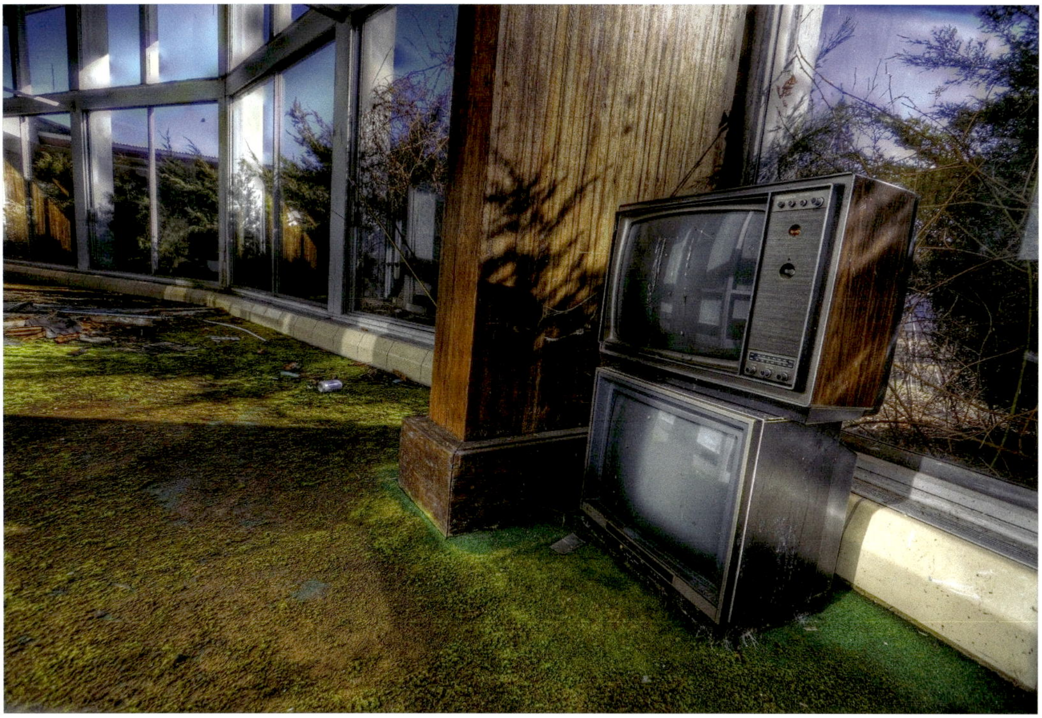

High-tech entertainment of the day at the abandoned Tamarack Lodge in Greenfield Park, New York. (*Photo: Andy Milford*)

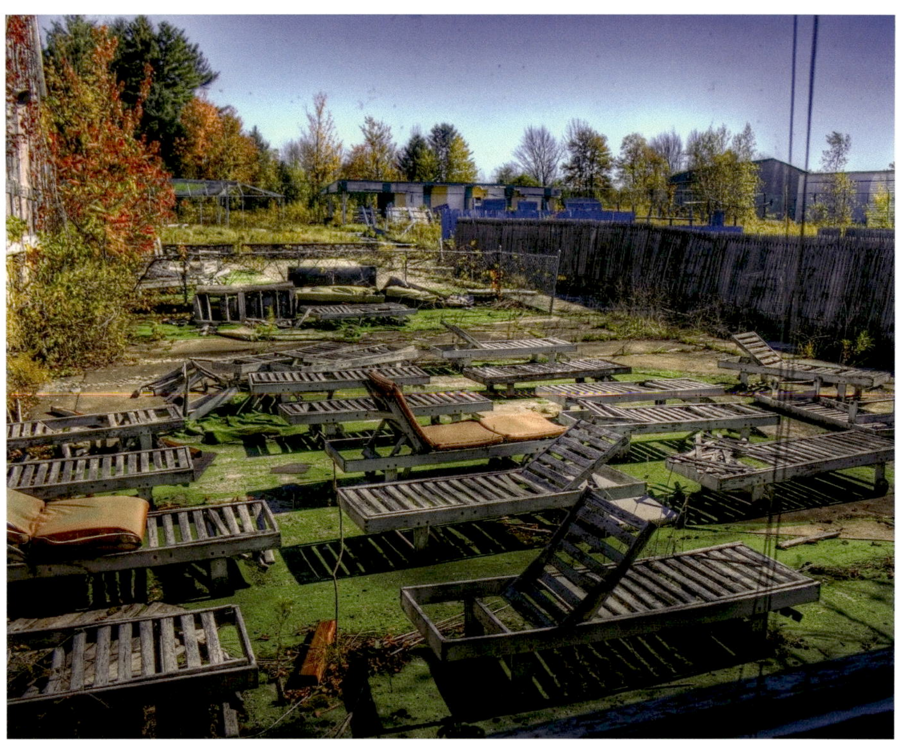

Outdoor chaises suggest the Tamarack once hosted many New York City sun worshippers. (*Photo: Andy Milford*)

Several factors contributed to the decline of the Borscht Belt resorts. One explanation was the changing vacation preferences of the American public. By the 1970s, air travel had become more affordable and Americans began exploring more distant and exotic destinations like Florida, California, and Europe. The social dynamics that had made the Borscht Belt a vibrant community also evolved. The younger generations of Jewish Americans were less inclined to seek out the culturally specific environment of the Catskills resorts. Additionally, societal shifts toward greater inclusivity meant that Jewish families faced fewer barriers in accessing other vacation destinations. The economic pressures on the Borscht Belt resorts increased over time. Maintaining large properties with extensive amenities required significant investment. As visitor numbers dwindled, many resorts struggled to stay profitable. The economic downturns of the late twentieth century further exacerbated these challenges, leading to closures and abandonments. These same social and economic realities also affected other nearby properties. Among the abandoned hotels in this part of the region are the Adler Hotel and others in Sharon Springs and the Cold Spring Hotel in Tannersville.

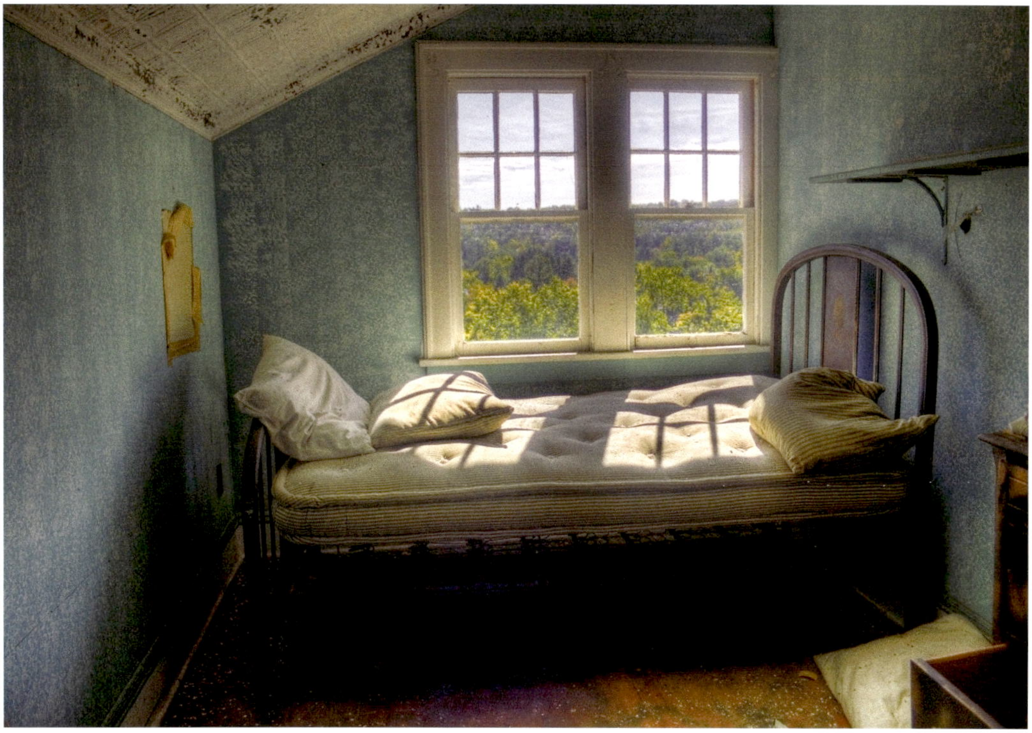

A modest room in the abandoned Adler Hotel in Sharon Springs, New York. (*Photo: Andy Milford*)

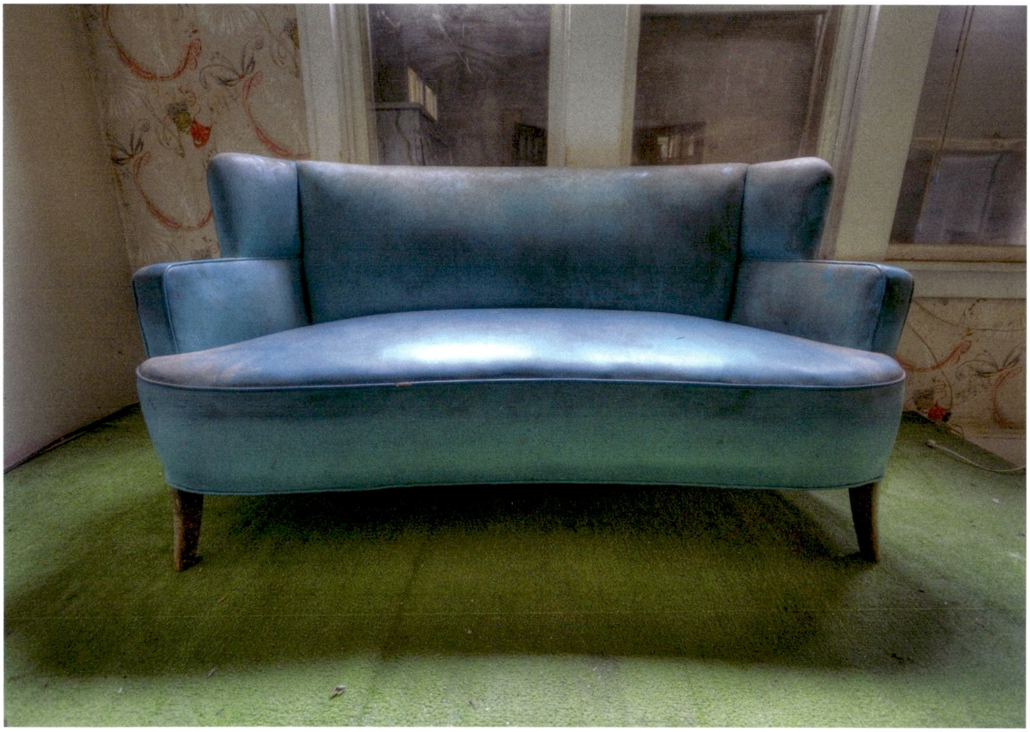

A bright blue couch was part of the allure at the Adler Hotel in Sharon Springs, New York. (*Photo: Andy Milford*)

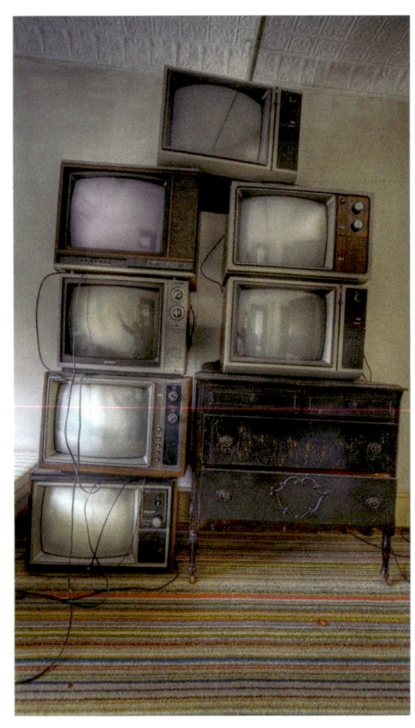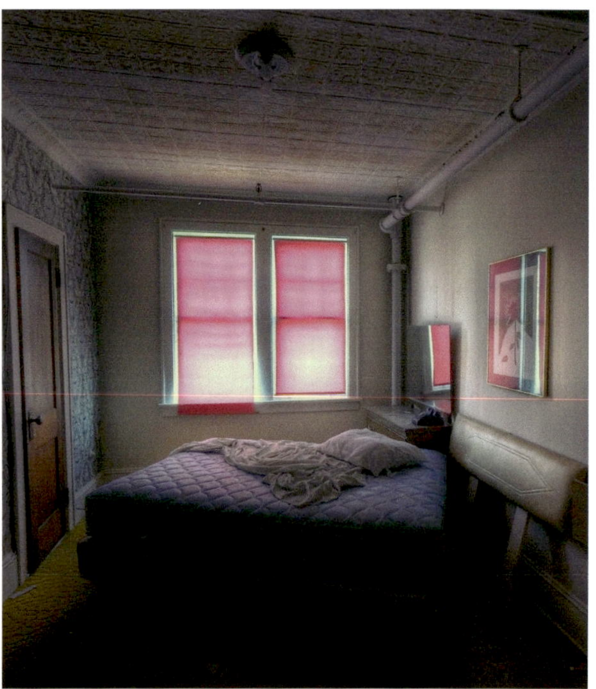

Above left: Sets upon sets of old school televisions accumulate at the Adler Hotel. (*Photo: Andy Milford*)

Above right: What remains of a stylish accommodation at the Adler Hotel. (*Photo: Andy Milford*)

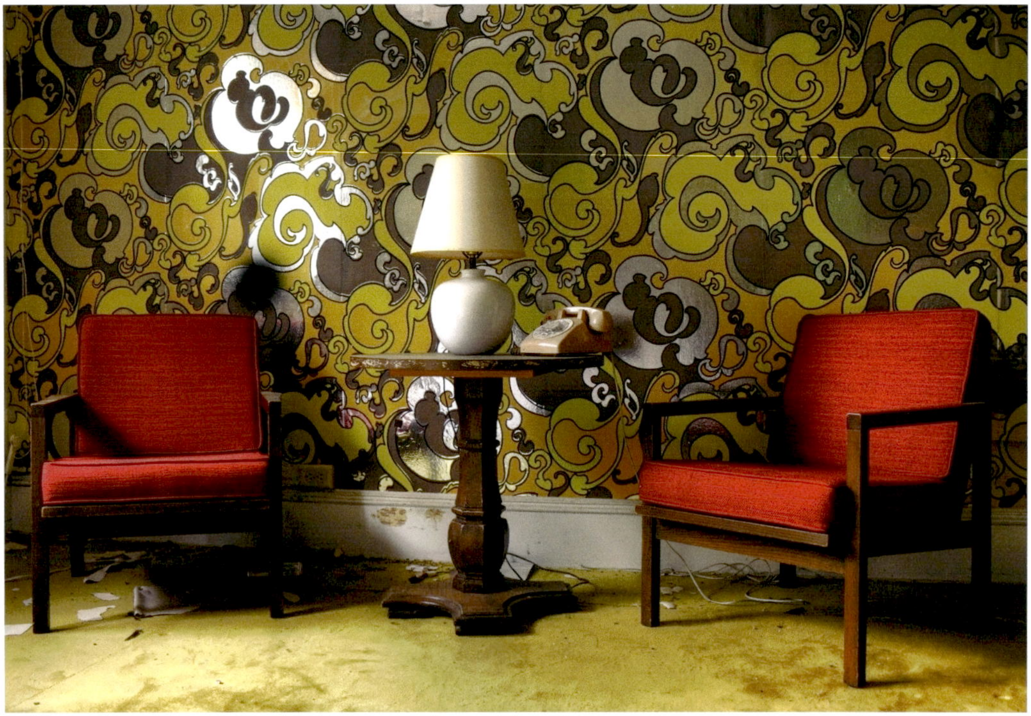

An inviting conversation grouping at the Adler Hotel. (*Photo: Andy Milford*)

5

INSTITUTIONS AND ASYLUMS

Entering an abandoned institution or asylum is like stepping into a ghost world, where the air is thick with melancholy and every noise seems like a cry in the night. Rooms that were once filled with the sounds of rehabilitation and recovery are now empty and silent. Furniture is overturned, papers are scattered across the floor, and the detritus of years of neglect litters the once-bustling hallways. Despite the decay and desolation, there is a strange beauty to be found in these abandoned institutions. Nature has begun to reclaim these buildings, with vines creeping through broken windows and trees growing up through the floors. Sunlight filters through the dust motes, casting a soft glow over the scene. Evidence of fires and vandalism is everywhere.

In 1874, the Middletown State Homeopathic Hospital (later Middletown Psychiatric Center) was established as "a state lunatic asylum for the care and treatment of the insane" and was described as little more than a single structure on a farm site, treating some sixty-nine patients. By 1974, it was "a city within a city," treating close to 1,400 and employing a staff of over 1,000. Believing that a healthy body was necessary for a healthy mind, treatment at Middletown included "baseball therapy" for male patients, and in 1888, Middletown State Hospital fielded a team of patients, staff, local amateurs and semi-pro players, called the "Asylums." The hospital earned a reputation for innovative treatments for mental illness, but this would not be enough to save Middletown. Like many other state mental hospitals, Middletown would struggle for survival. While some buildings are still in use, many are now vacant.

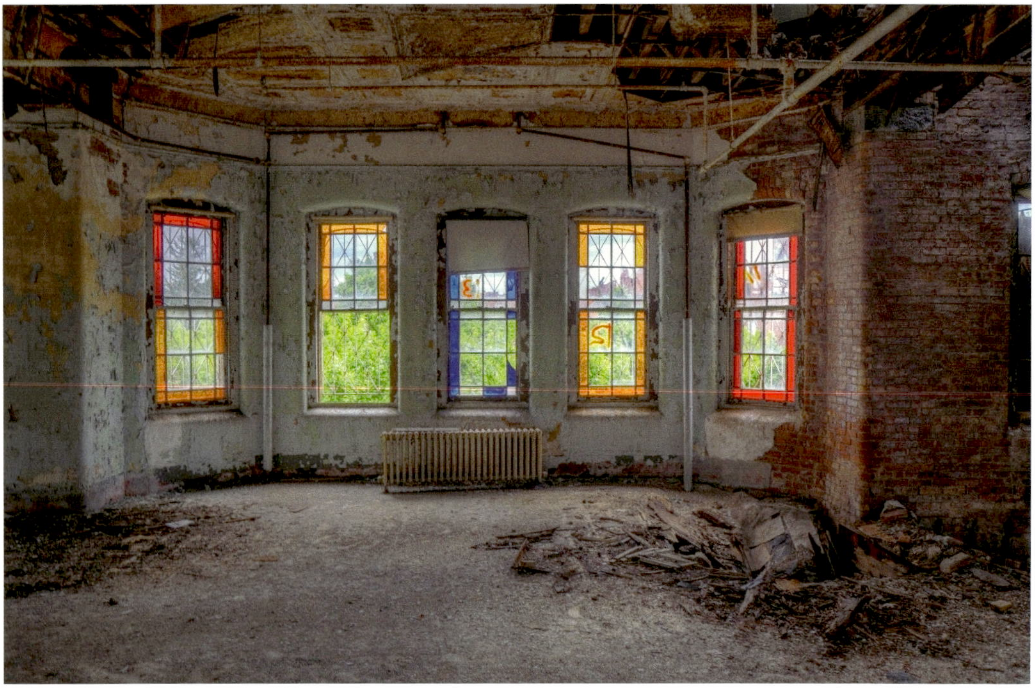

Broken stained-glass windows at Hudson River State Hospital suggest years of neglect and abandonment. (*Photo: Andy Milford*)

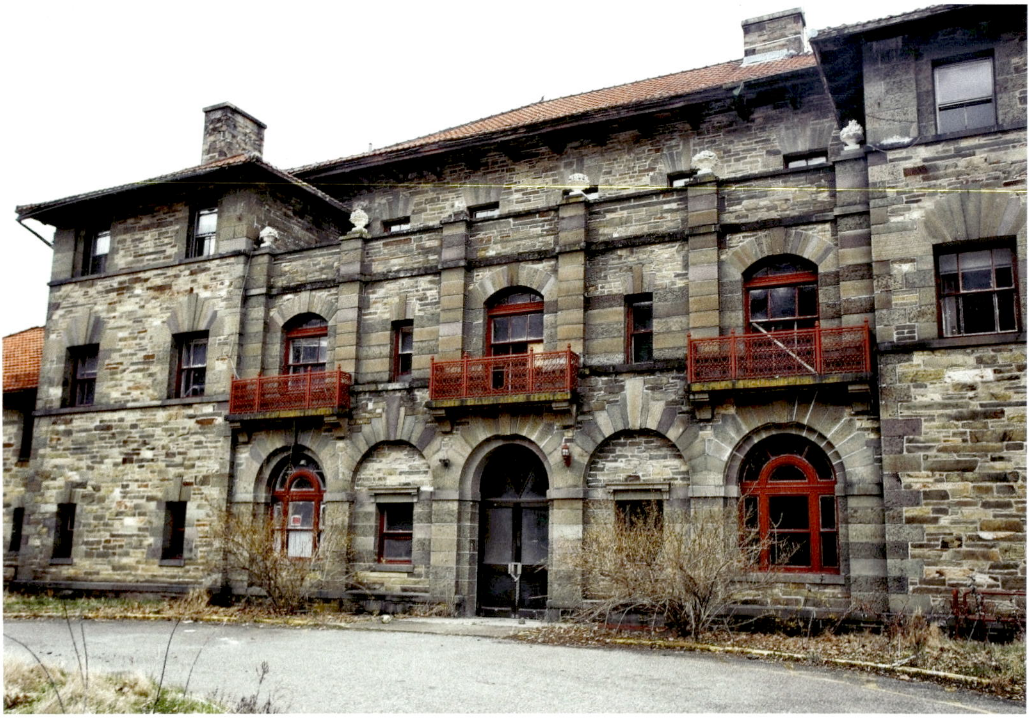

The former Miller estate in Rhinebeck, New York had several lives—as a private residence, health resort, and later a substance abuse facility. (*Photo: Liz Cooke*)

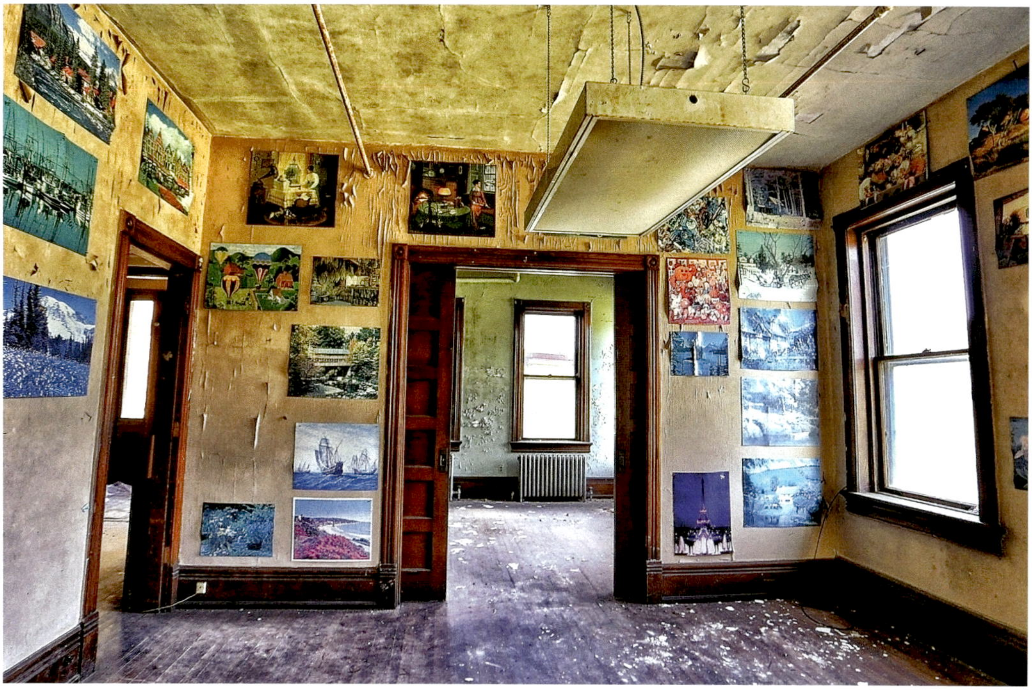

The "puzzle room" at the Dutchess County Poorhouse in Millbrook, New York. The Poorhouse was built in 1864 and closed in 1998. (*Photo: Liz Cooke*)

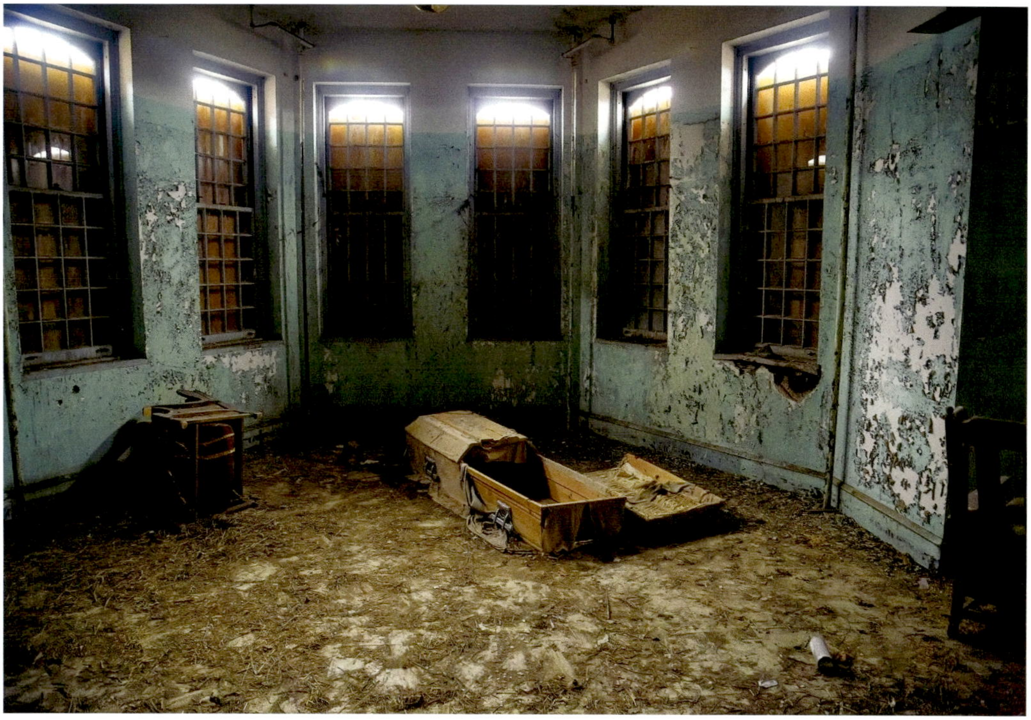

A coffin lays bereft and forgotten in an abandoned building at Middletown Psychiatric Center. (*Photo: Andy Milford*)

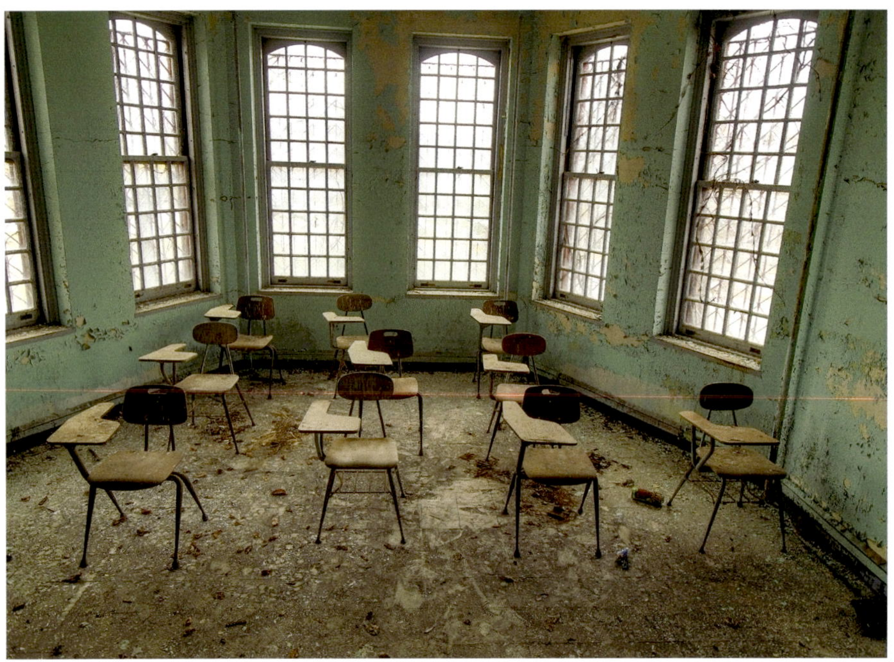

Chairs arranged as if for a class tell an eerie story at Middletown Psychiatric Center. (*Photo: Andy Milford*)

Opened in 1871 as the Hudson River State Hospital for the Insane, Hudson River Psychiatric Center is located in Poughkeepsie, New York. The landmark property was notable for its High Victorian Gothic architecture, especially its majestic administration building, designed by Frederick Clarke Withers according to the Kirkbride plan, a design for mental asylums based on the principle of moral treatment. Hudson River State Hospital was the first institution to use the Kirkbride plan, creating a sense of grandeur, privacy, and curative environments for patients. The grounds were designed by Central Park landscape designers Calvert Vaux and Frederick Law Olmstead. Construction began in 1868, and when it opened three years later, it housed forty patients. By 1952, that number grew to almost 6,000; and by 1976, as psychiatric treatment moved to medication management and community-centered care, the number of patients dwindled to fewer than 1,800.

Hudson River State Hospital closed in 2003. Much of the male wing was destroyed in a 2007 fire. After many years on the market and several more fires, Hudson River State Hospital was sold in November 2013. Development is well under way with plans to reuse several of the structures, including the administration building, theater and chapel. A busy Shop-Rite, Starbucks, and Chipotle are tenants of the new development.

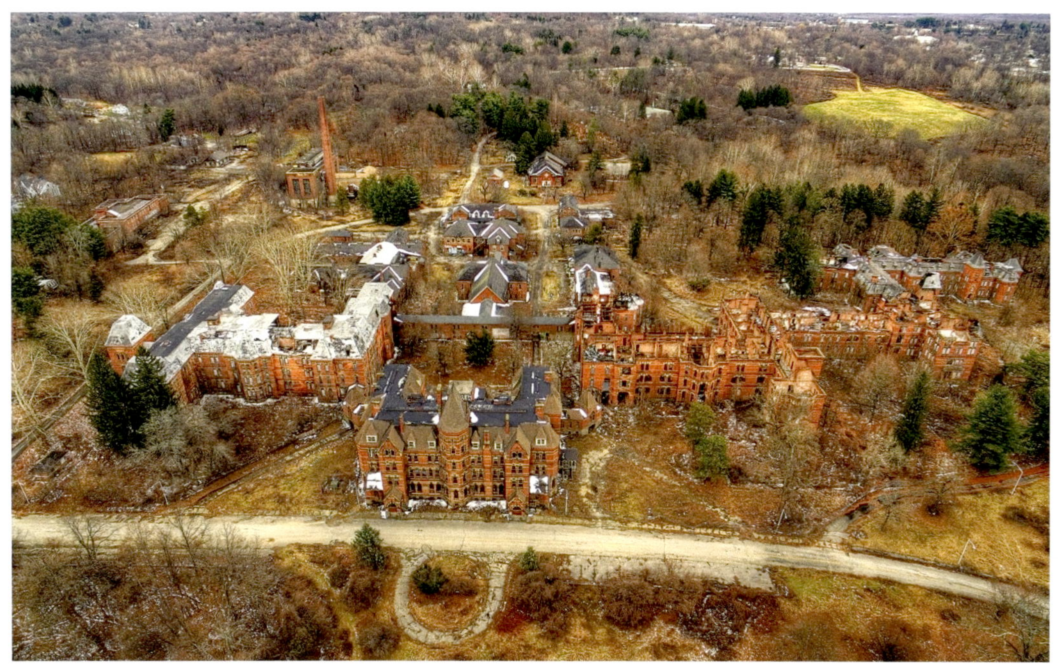

An aerial view of the 296-acre campus of the Hudson River State Hospital in Poughkeepsie, New York. Many of these buildings have since been demolished to make way for a mixed-use development. (*Photo: Andy Milford*)

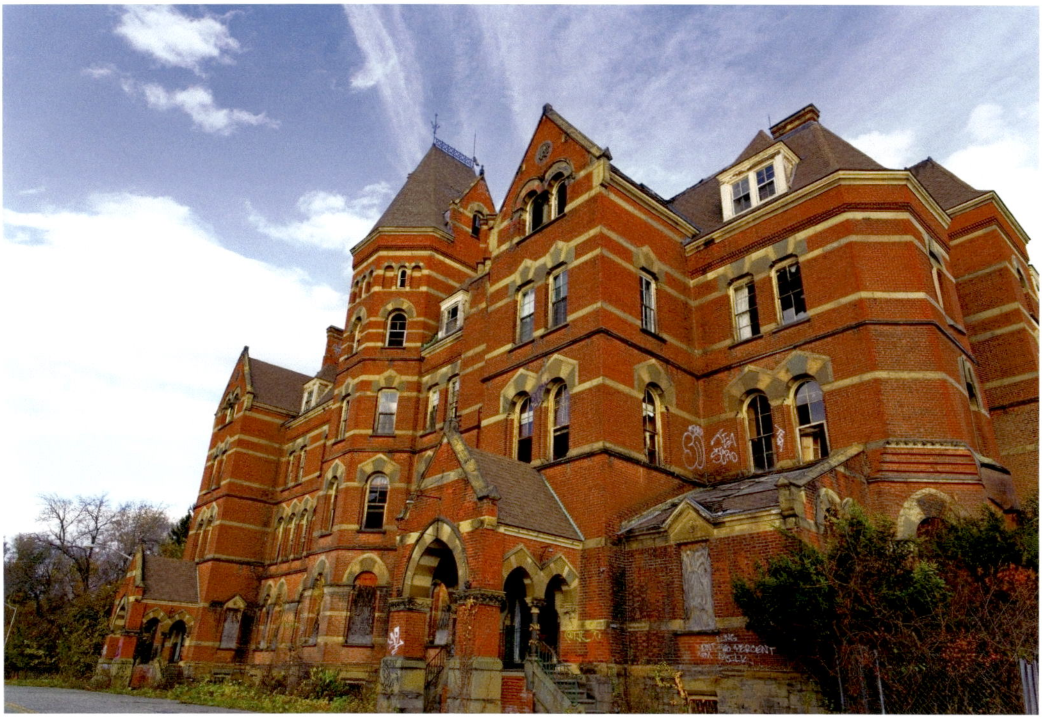

The administration building at Hudson River State Hospital in Poughkeepsie, New York, was built in the high Victorian Gothic architecture style of the day. Development plans call for the building to be reborn as a hotel. (*Photo: Liz Cooke*)

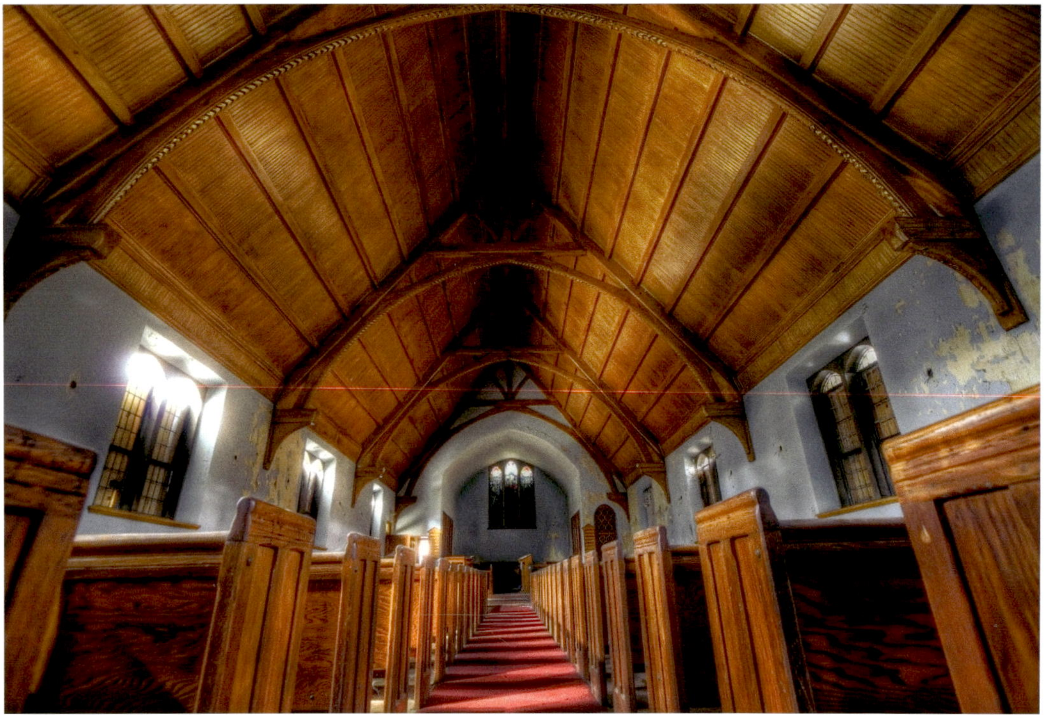

The Avery Chapel at Hudson River State Hospital is expected to be saved and repurposed as development proceeds at the site. (*Photo: Andy Milford*)

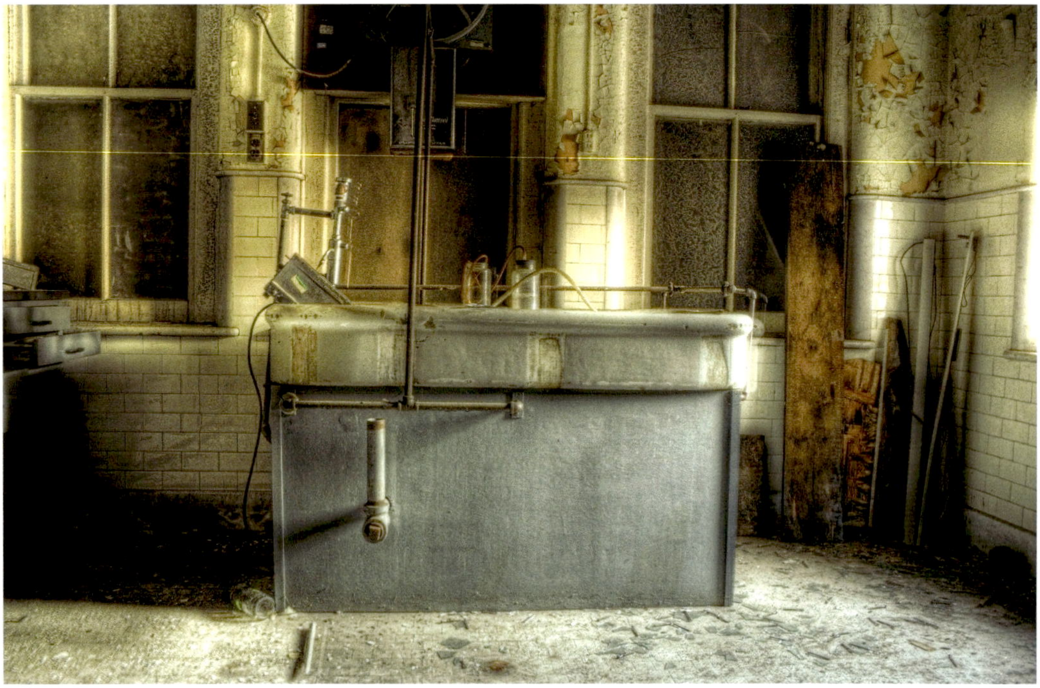

An autopsy table in the morgue building at Hudson River State Hospital in Poughkeepsie, New York. (*Photo: Andy Milford*)

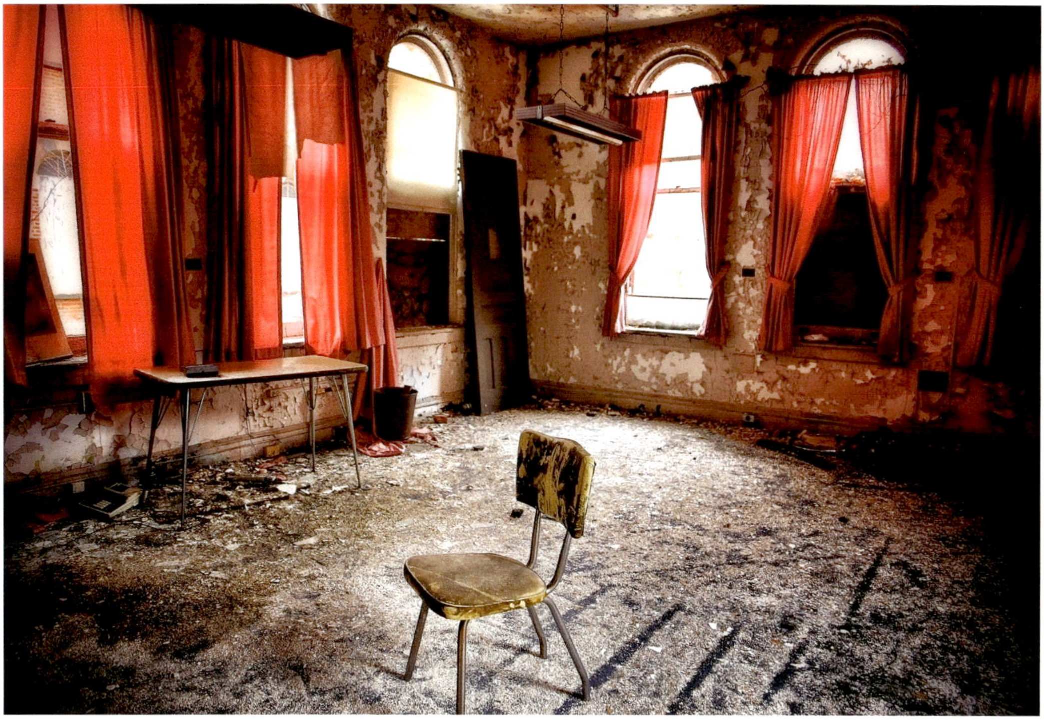

The "art room" was used for patient rehabilitation in the sprawling administration building at Hudson River State Hospital. (*Photo: Liz Cooke*)

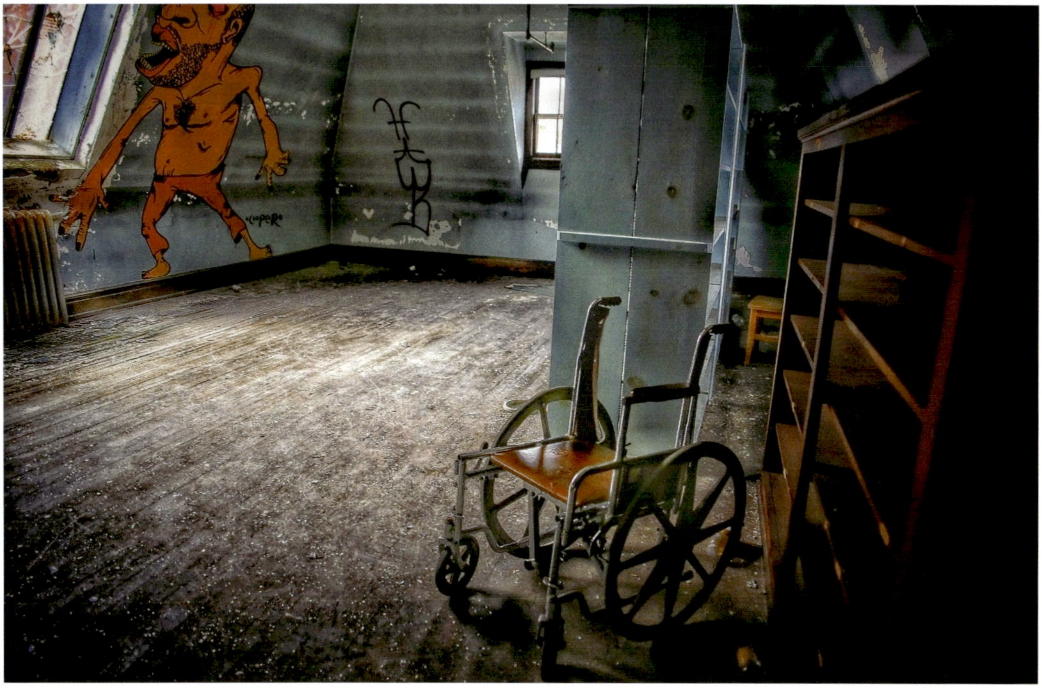

An attic room in the administration building has an impressive piece of street art. (*Photo: Liz Cooke*)

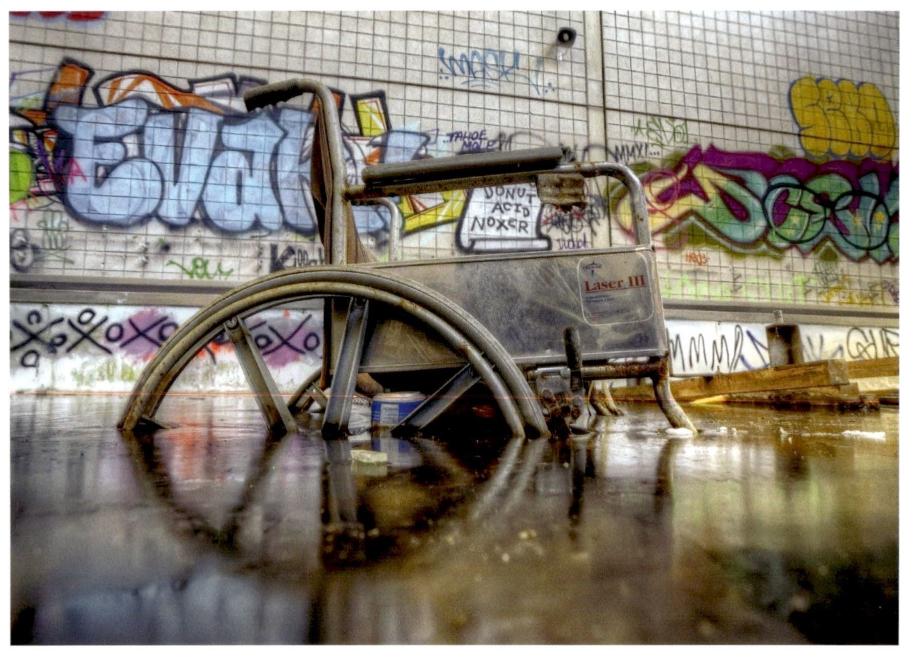

A wheelchair is found embedded in ice in the swimming pool of the now-demolished Recreation Center at Hudson River State Hospital. (*Photo: Andy Milford*)

Letchworth Village, located in Thiells, New York, was an institution for the developmentally disabled that opened in 1911 and closed in 1996. The village was initially designed to provide a more humane and home-like environment for its residents. However, over the decades, it gained a notorious reputation for overcrowding, neglect, and inhumane treatment of patients. Investigations and exposés, particularly in the 1970s, revealed the dire conditions and led to widespread public outrage, contributing to reforms in the care of the developmentally disabled. Today, much of the site stands abandoned, a haunting reminder of its troubled past.

Craig House Hospital in Beacon, New York, was a renowned private psychiatric facility known for its progressive treatment methods and distinguished clientele. Founded by Dr. Clarence Slocum and his wife, the hospital operated from a grand estate, providing a serene and luxurious environment intended to promote mental healing. Craig House became famous for treating several high-profile patients, including author Zelda Fitzgerald, actress Frances Farmer, and Frances Ford Seymour (Jane Fonda's mother) who committed suicide at Craig House in 1950. By the late twentieth century, changes in psychiatric care practices, along with financial difficulties, led to the hospital's closure. Since then, the Craig House estate had remained largely vacant. In 2024, Mirbeau Spa took ownership of the property and has begun an extensive renovation.

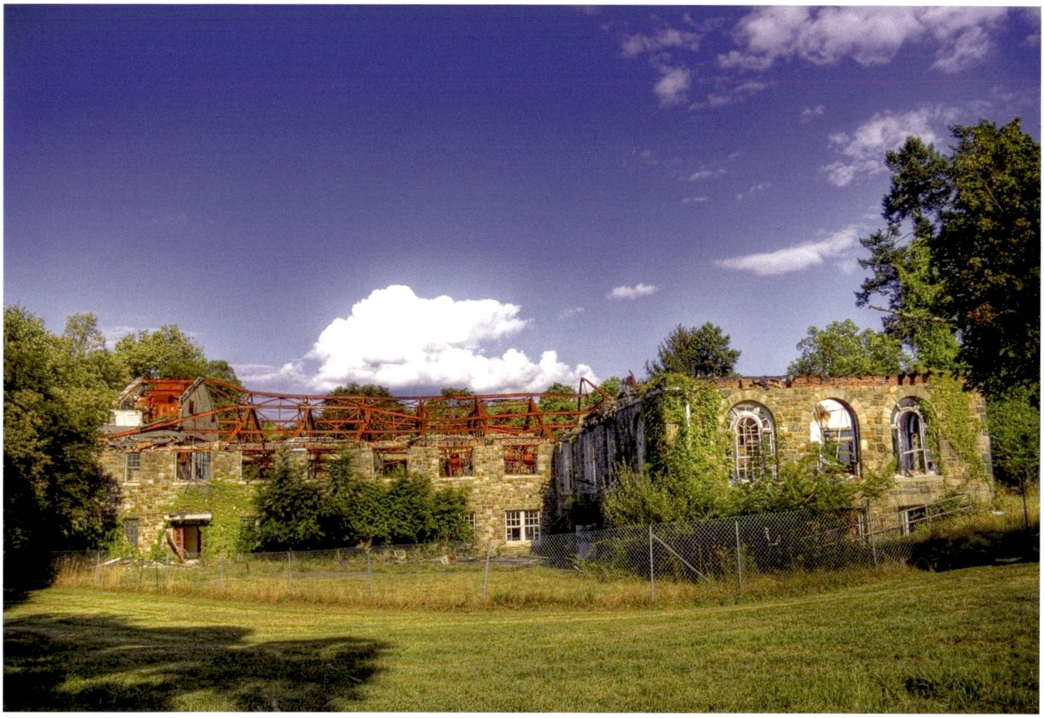

The grounds of Letchworth Village in Thiells, New York, are open to the public. (*Photo: Andy Milford*)

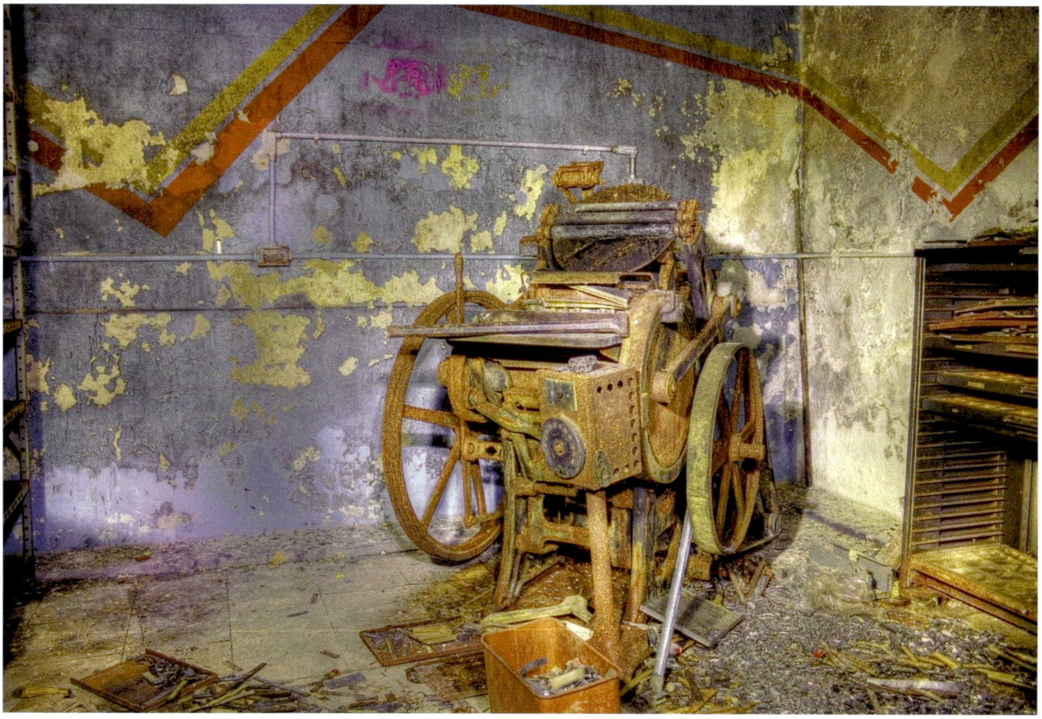

Rusting machinery is a reminder of busier times at Letchworth Village, a facility that earned a tragic reputation for poor treatment of its vulnerable residents. (*Photo: Andy Milford*)

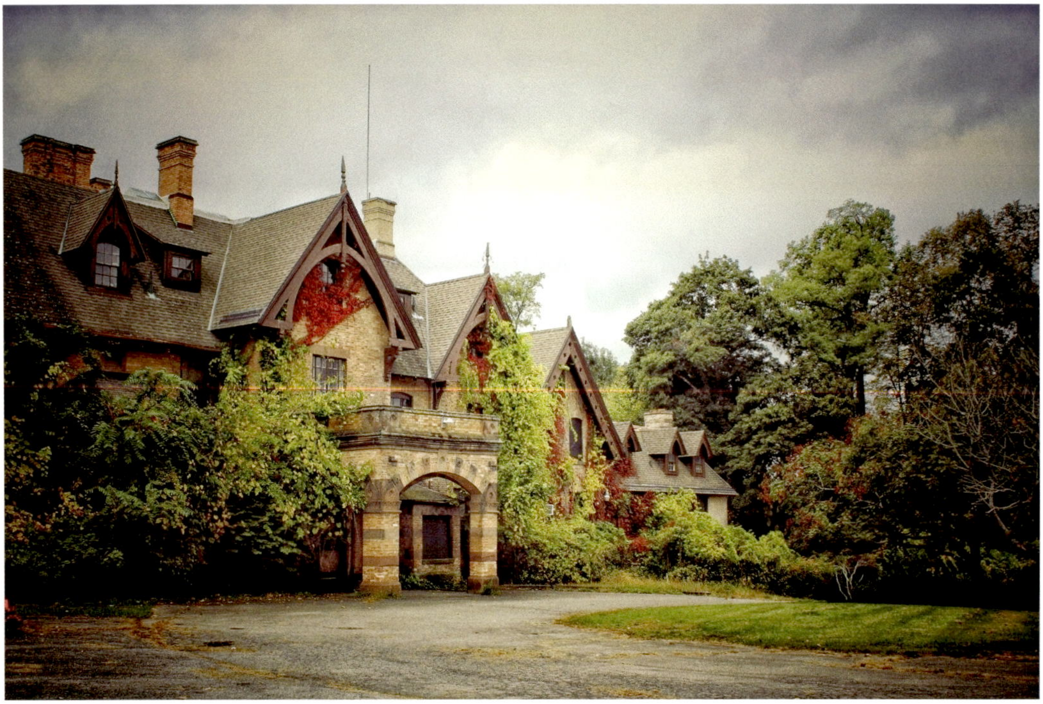

Craig House Hospital in Beacon, New York, was once known for innovative treatments for mental illness and a celebrated clientele. (*Photo: Liz Cooke*)

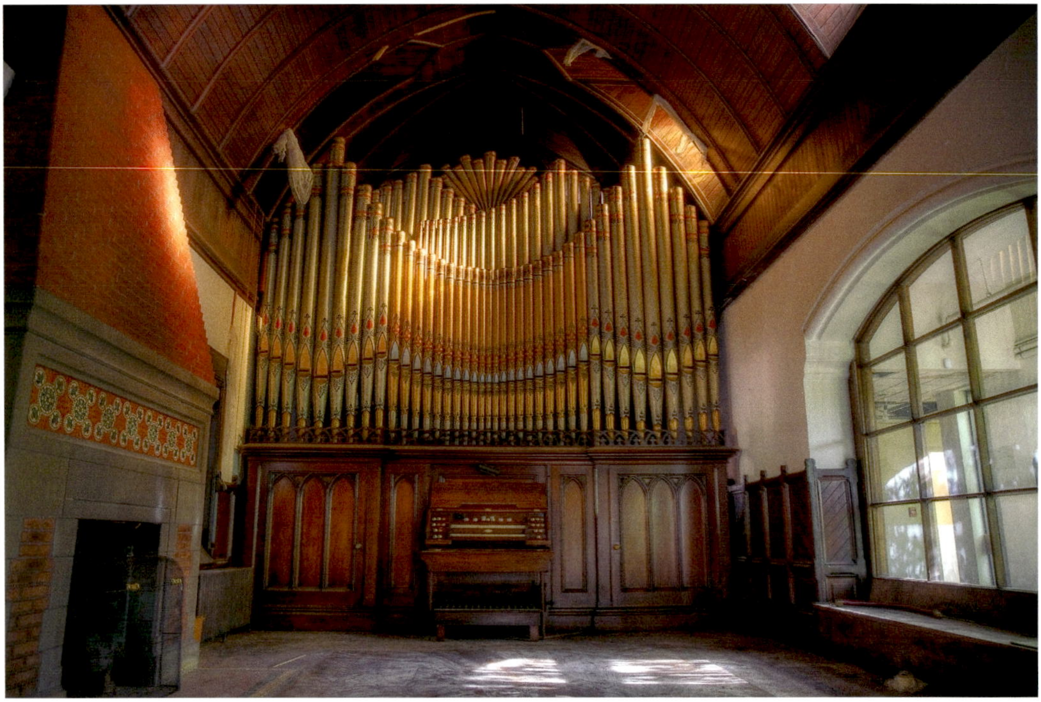

A grand organ at Craig House Hospital bespeaks a level of elegance unknown at other psychiatric institutions. (*Photo: Andy Milford*)

Camp LaGuardia, located in Chester, New York, was established in 1934 as a New York City-operated homeless shelter for men. Spanning over 1,000 acres, the camp provided refuge and rehabilitation services for unhoused men, offering them shelter, meals, and work opportunities in a rural setting far from urban stresses. For several decades, Camp LaGuardia played a crucial role in addressing homelessness, housing thousands of men and serving as a model for similar initiatives. However, as the years progressed and social policies evolved, the camp's relevance diminished. In 2007, the facility was closed. Since then, the property has had various proposals for redevelopment, but much of the site remains abandoned.

Throughout the Hudson Valley there are numerous facilities designed to care for the most vulnerable members of society. Among these were the Dutchess County Poorhouse in Millbrook, New York, a tuberculosis sanatorium, a private mansion reborn as a health resort (later a substance abuse facility) and several convalescent homes. Among these is the former Brandywine Estate (later the Briarcliff Manor Center). Here you will find empty rooms with empty closets and extravagantly tiled bathrooms outfitted with handrails and other accommodations for the elderly and infirm. The Brandywine property closed when a modern facility was built nearby.

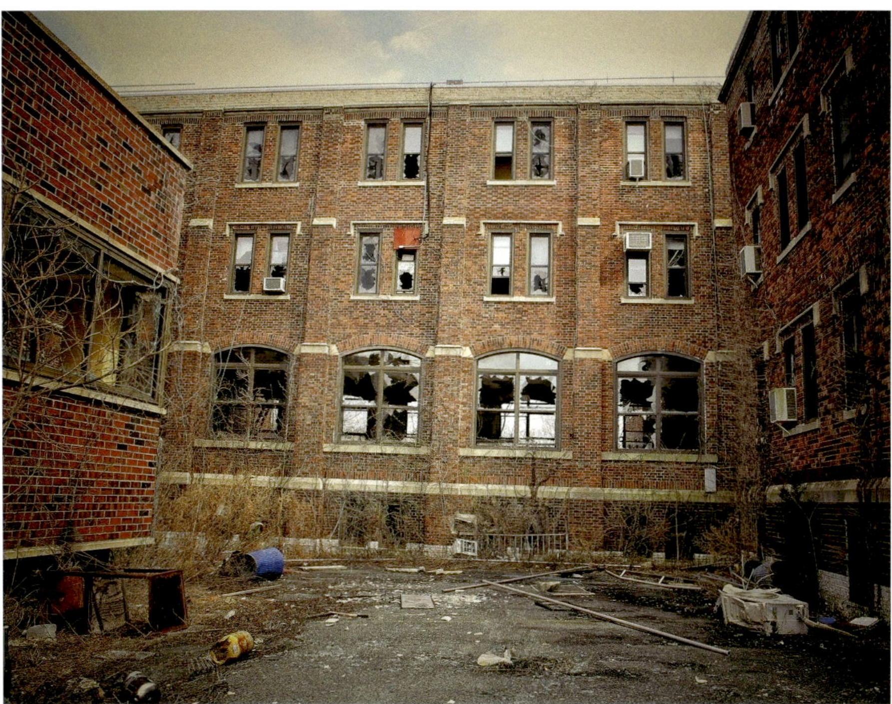

Exterior shot at Camp LaGuardia in Chester, New York, shows the effects of neglect and abandonment at this former homeless shelter for men. (*Photo: Liz Cooke*)

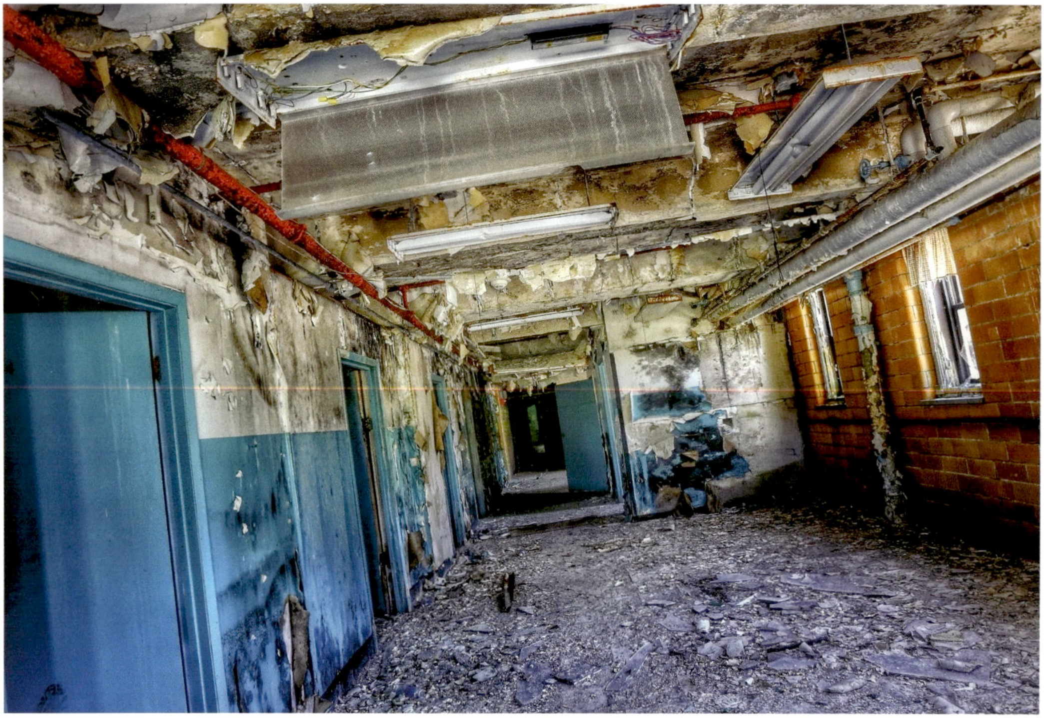

Rooms that once housed thousands of New York City's homeless men are empty and abandoned at Camp LaGuardia in Chester, New York. (*Photo: Liz Cooke*)

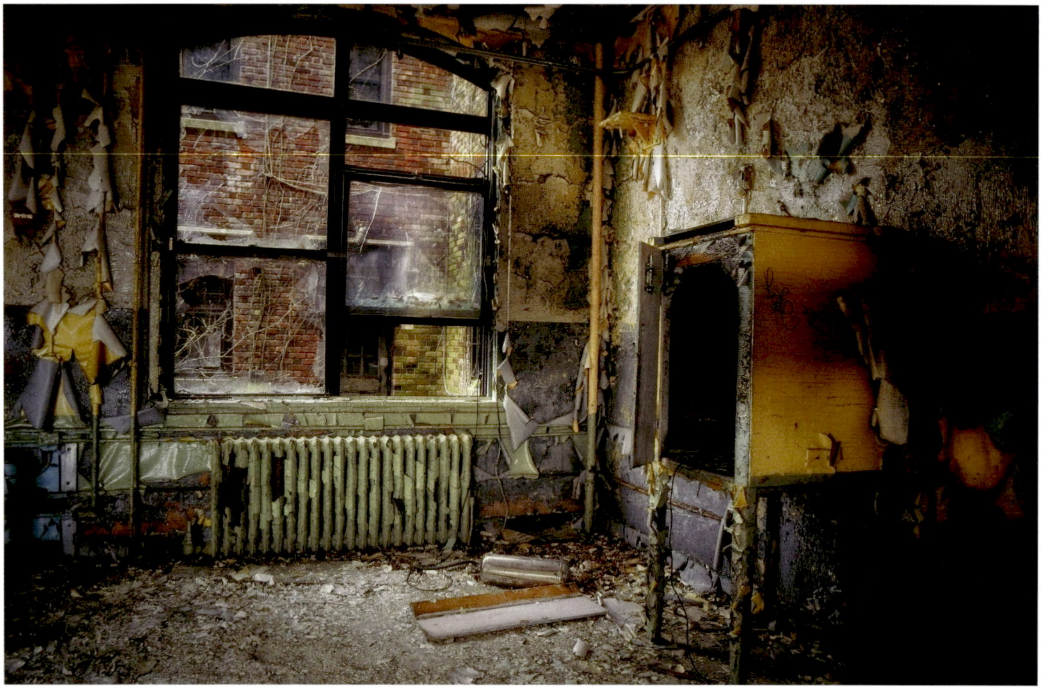

Broken glass, peeling paint and weathered walls tell a story of abandonment at Camp LaGuardia in Chester, New York. (*Photo: Liz Cooke*)

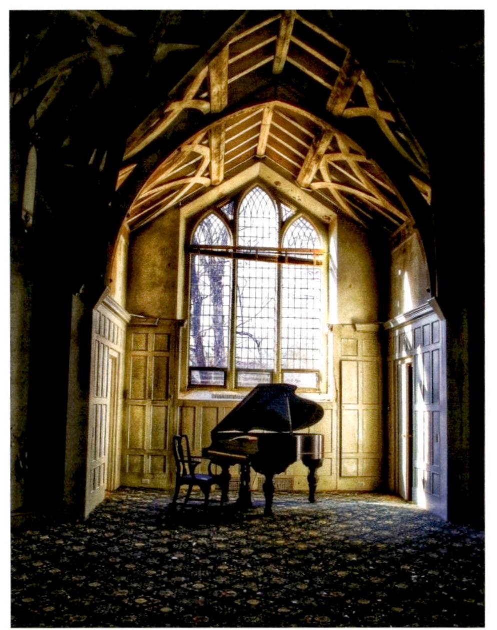 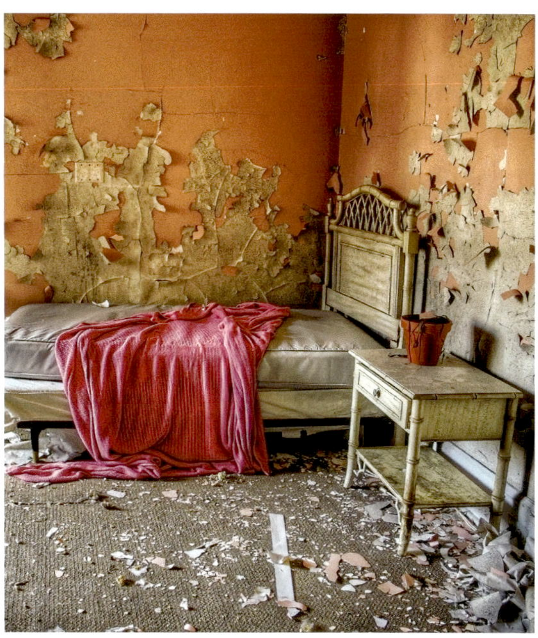

Above left: The music room in what was once known as the Brandywine Estate. (*Photo: Liz Cooke*)

Above right: Residents of the Briarcliff Manor Center for Rehabilitation and Nursing enjoyed comfortable rooms. (*Photo: Liz Cooke*)

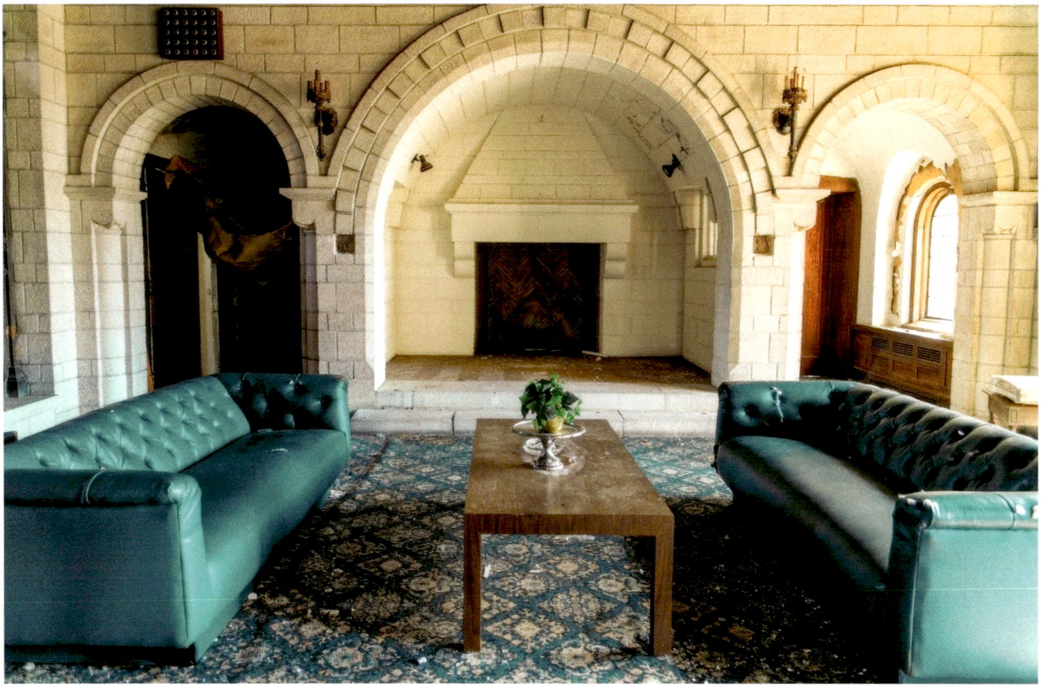

The "Great Room" at the Briarcliff Manor Center boasts arched windows and a massive fireplace. (*Photo: Liz Cooke*)

6

INDUSTRY AND TRANSPORTATION

The Hudson Valley has a rich industrial history that reflects broader economic and technological trends in the United States. Over the centuries, industries in the region evolved significantly, responding to changing demands, resources, and innovations. This evolution brought success to some industries that were able to adapt and the end of other industries, leaving little more than abandoned buildings and fading memories. It is worth noting that many of the structures associated with these bygone industries had ornate architectural details and a level of workmanship seldom seen in modern buildings. While major industries like ice harvesting and brickmaking may have become obsolete or at least less prevalent, many newer industries as well as independent retailers, restaurants and other small businesses remain. But even these newer operations cannot always avoid the fate of abandonment. And like many regions in the nation, the Hudson Valley is not without a small handful of abandoned malls.

Cement production gained prominence in the Hudson Valley during the nineteenth century. The discovery of high-quality limestone deposits led to the establishment of several cement plants. Rosendale cement became famous for its durability and was used in the construction of the Brooklyn Bridge and the Statue of Liberty's pedestal. This industry experienced a decline in the early twentieth century as Portland cement, which was easier to produce and use, became more popular. Smaller cement factories and the massive industrial grounds of Cementon in Catskill, New York, are all that remain of a once thriving industry.

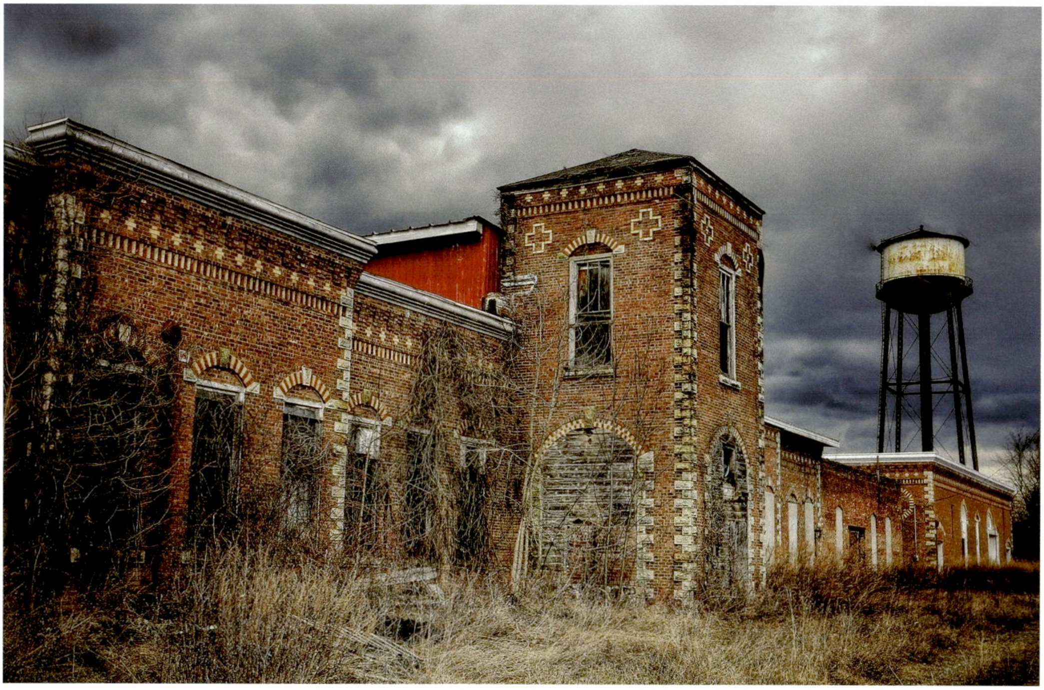

Located in Wallkill, New York, the Borden Condensed Milk Factory was innovative for its time, offering new methods of food preservation. Abandoned in the mid-twentieth century, plans are under way for an adaptive reuse of the property, including an upscale hotel to be called "Moo." (*Photo: Liz Cooke*)

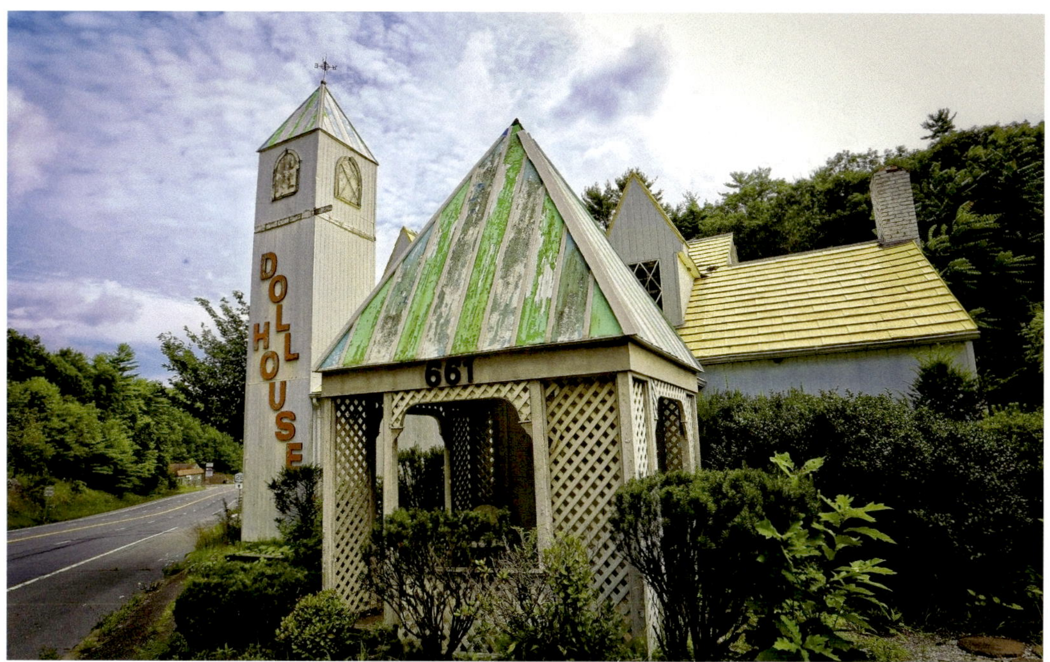

The Doll House, outside Kingston, New York, had a slightly suggestive name but was actually quite innocent, selling nothing but dolls and dollhouses. It was abandoned in the mid-1990s and has been reborn as a window treatment franchise. (*Photo: Liz Cooke*)

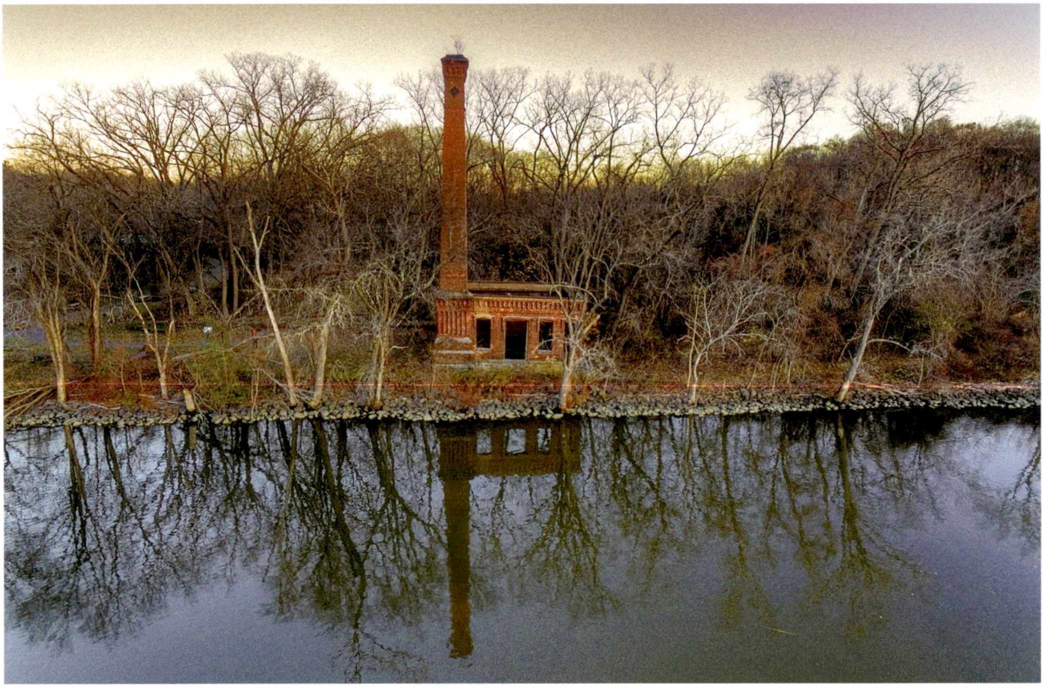

The R. and W. Scott Ice Company Icehouse and Powerhouse in Stuyvesant, New York, is a well-preserved piece of Hudson Valley history and one of a very few remaining icehouses in the Hudson Valley. (*Photo: Andy Milford*)

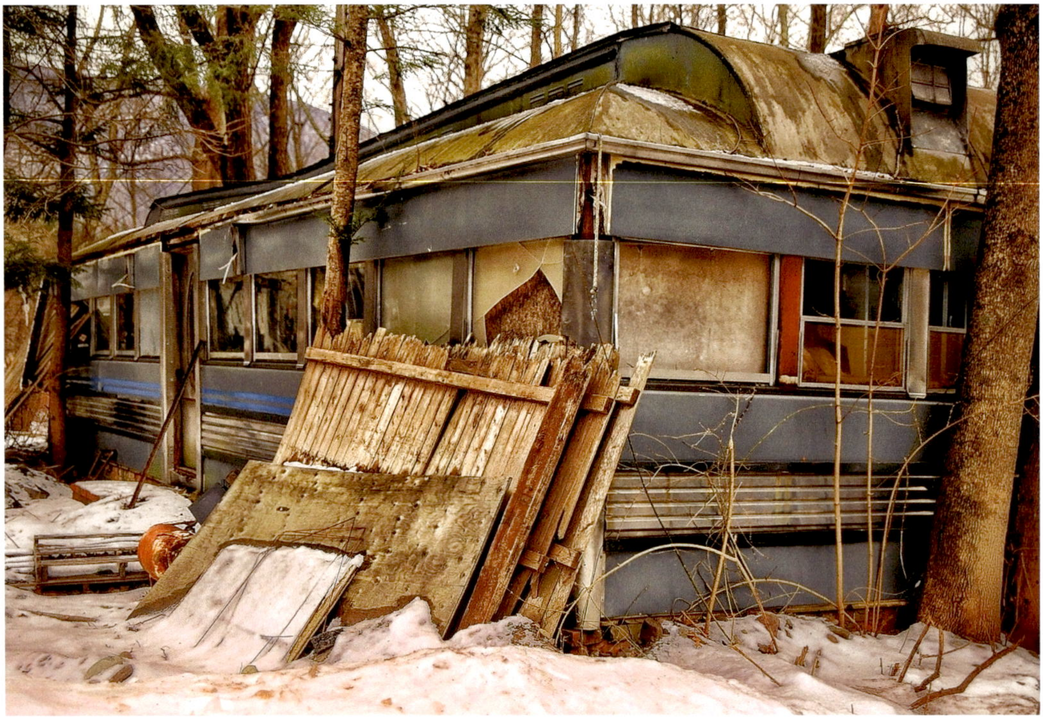

Elmer's Diner in Phoenicia, New York, was once the place to go for simple home cooking. The old-fashioned dining car was abandoned for many years and used for storage. (*Photo: Liz Cooke*)

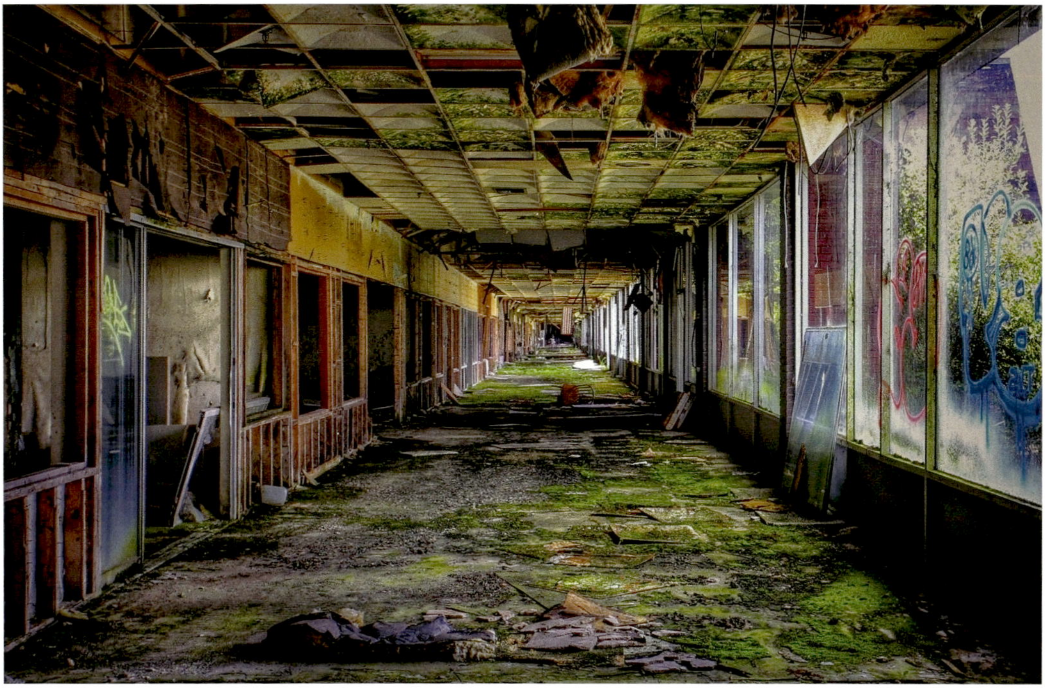

The abandoned Apollo Mall, located in Monticello, New York, was once Sullivan County's only indoor mall, complete with a mini-golf course and movie theater. Its last retail tenant moved out in 2003. (*Photo: Liz Cooke*)

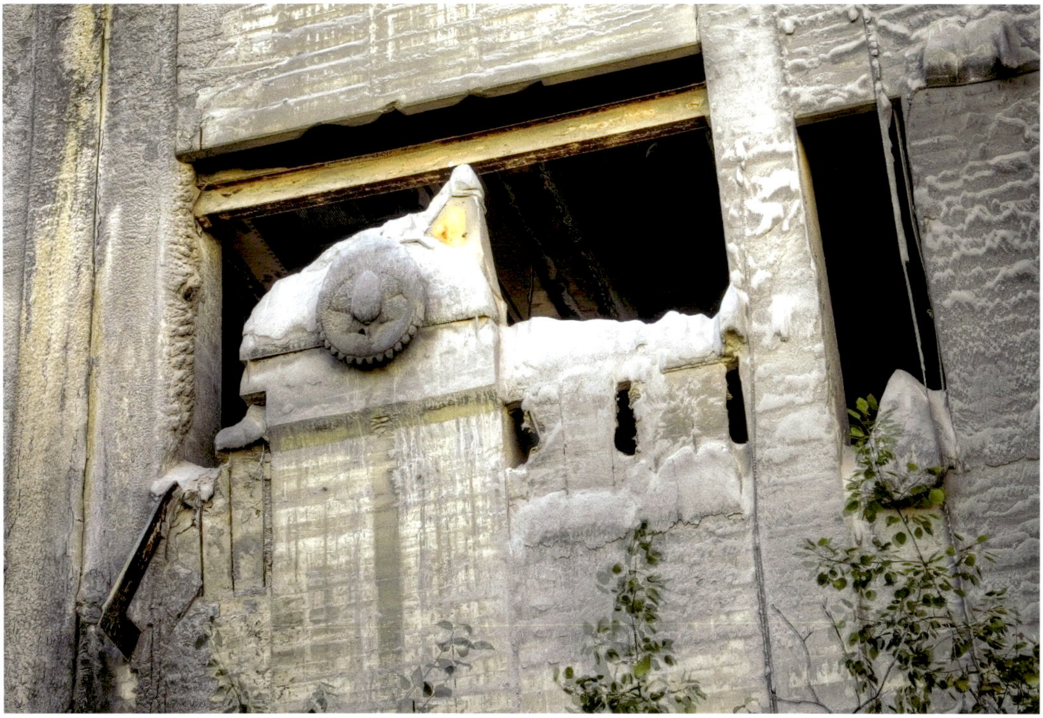

A thick layer of encrusted cement covers a once-needed fan at this abandoned cement plant in Cementon, New York. (*Photo: Andy Milford*)

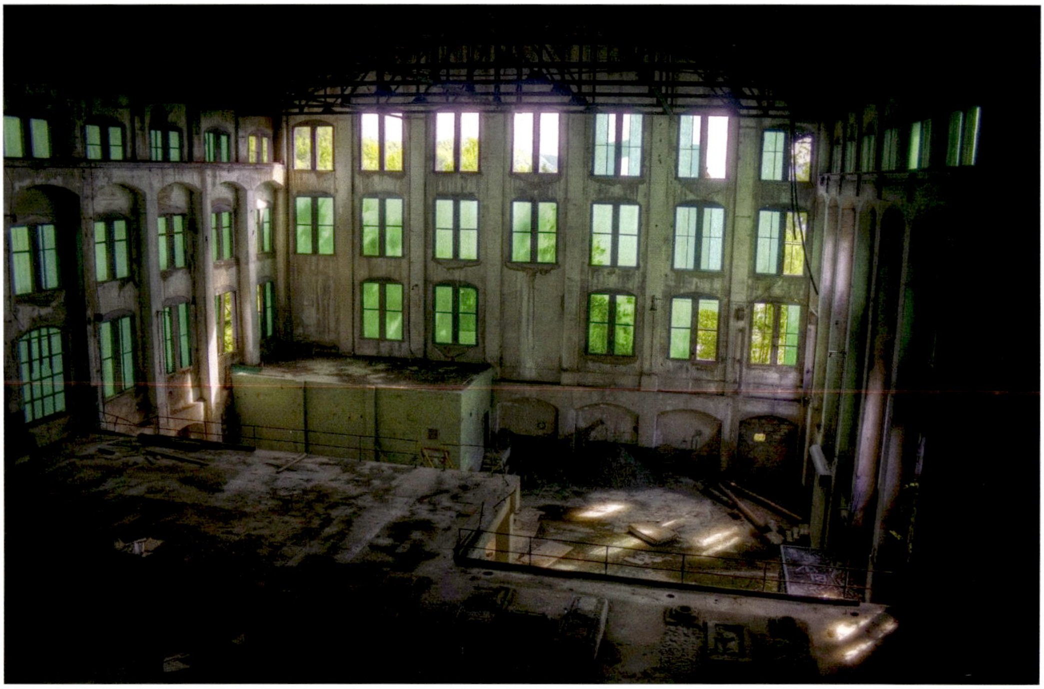

Light pours through the green-shrouded windows at the abandoned Hudson Cement plant in Hudson, New York. (*Photo: Andy Milford*)

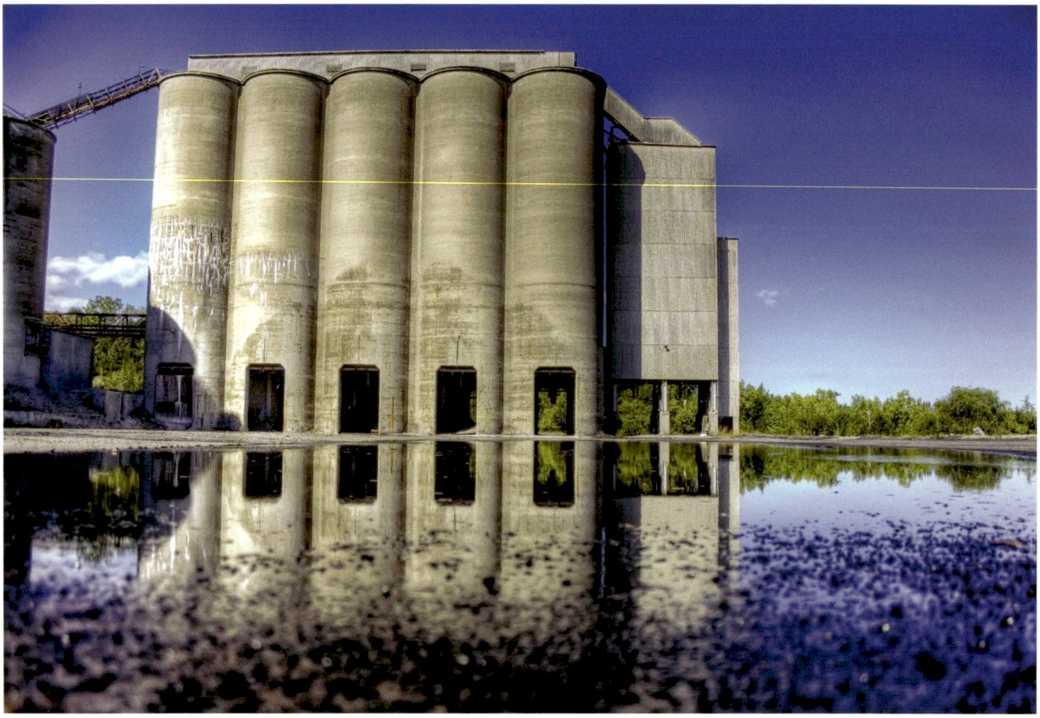

Cement silos are designed to make cement production more efficient. These abandoned silos are seen in the hamlet of Cementon, New York. (*Photo: Andy Milford*)

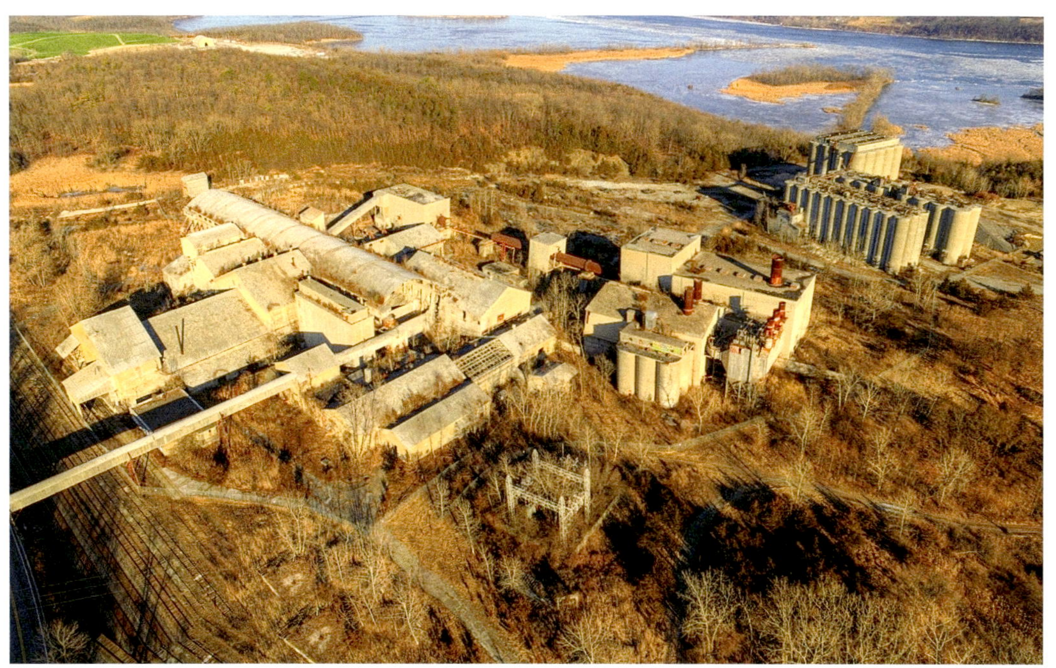

Located in Catskill, New York, the hamlet of Cementon once housed the massive Lehigh Cement plant. Cementon had housing for workers, schools, grocery stores, an icehouse, and a post office. The plant operated until 1982. (*Photo: Andy Milford*)

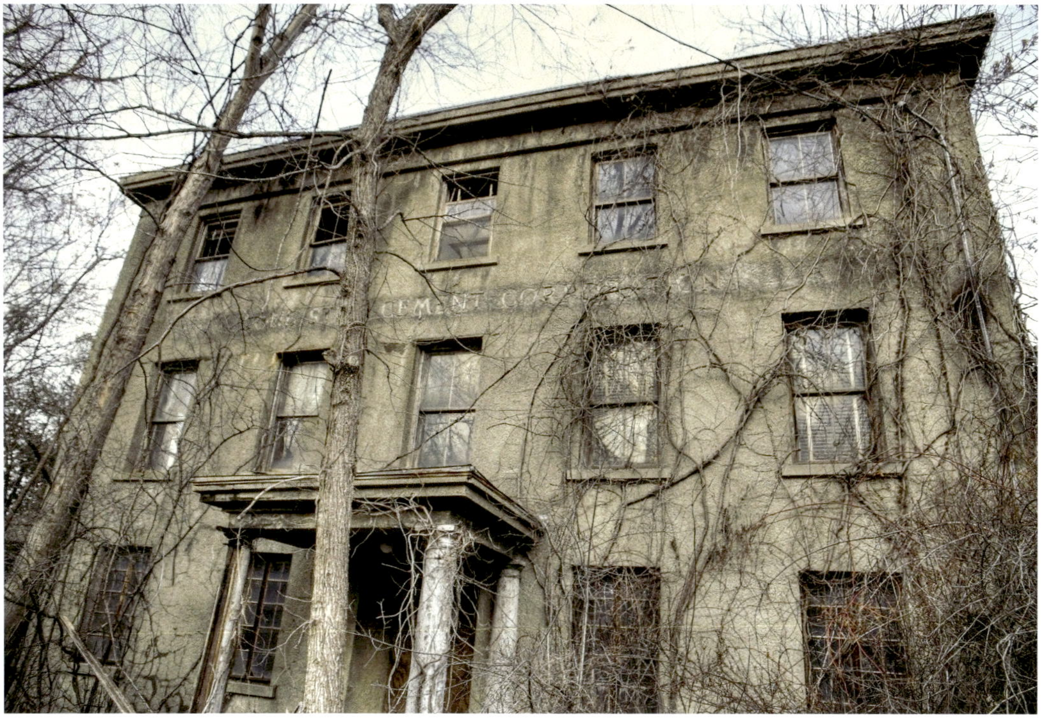

The remains of the Lone Star Cement Plant in Hudson, New York. Cement production, once a vital industry, faded in part due to rising wages as well as transportation and materials costs. (*Photo: Liz Cooke*)

The twentieth century saw the Hudson Valley's industrial landscape undergo significant changes. While some traditional industries declined, new ones emerged. A fully abandoned (and totally demolished) site is that of the Nepera Chemical Plant in Harriman, New York, a bulk producer of pharmaceutical and industrial chemicals. Opened in 1942 and closed in 2005, the site used a second facility 25 miles away in Hamptonburgh, New York, for wastewater disposal. Between 1953 and 1967, the plant dumped thousands of tons of chemical and sewage waste in a series of "lagoons," located scarily close to the town of Maybrook's water supply. When it was determined the lagoons were toxic, they were backfilled with dirt. Over the decades, the plant faced increasing scrutiny due to environmental concerns, including chemical spills and pollution. By the early 2000s, these environmental issues, coupled with changing economic conditions and regulatory pressures, resulted in the plant's closure. The Nepera Chemical Plant was placed on the Superfund's Priority List in 1986.

The Yonkers Power Station (also called the Glenwood Power Plant) in Yonkers, New York, was a striking example of early twentieth-century industrial architecture but is now in a state of almost total deterioration. Built in 1906, this once-state-of-the-art facility played a crucial role in powering the electrification of the New

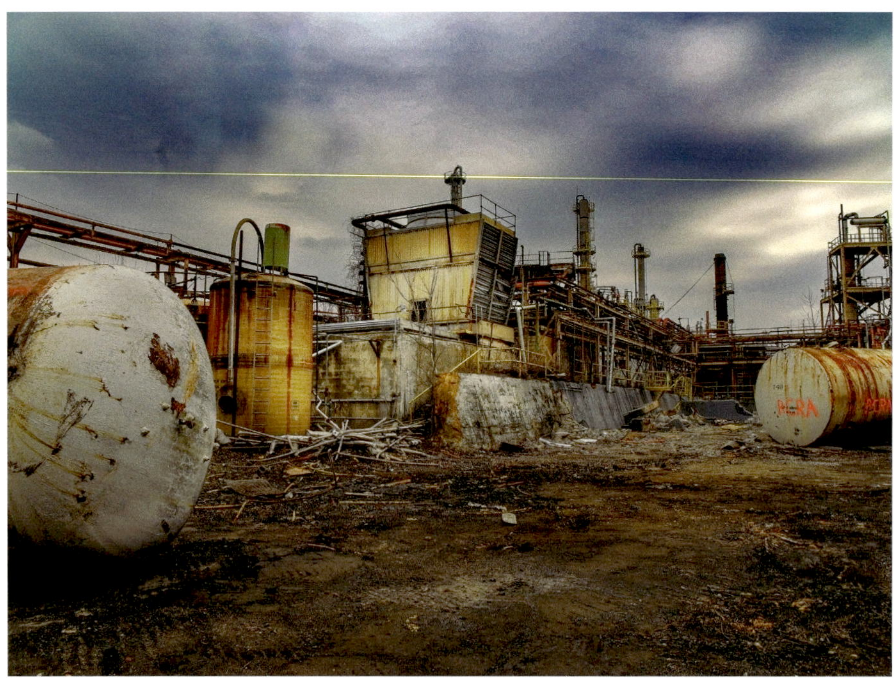

The Nepera Chemical Plant in Harriman, New York, was a bulk producer of industrial and pharmaceutical chemicals. Wastewater from the plant was sent to a second plant 25 miles away and dumped in a series of "lagoons." Environmental concerns caused the closing of the plant in 2005. (*Photo: Liz Cooke*)

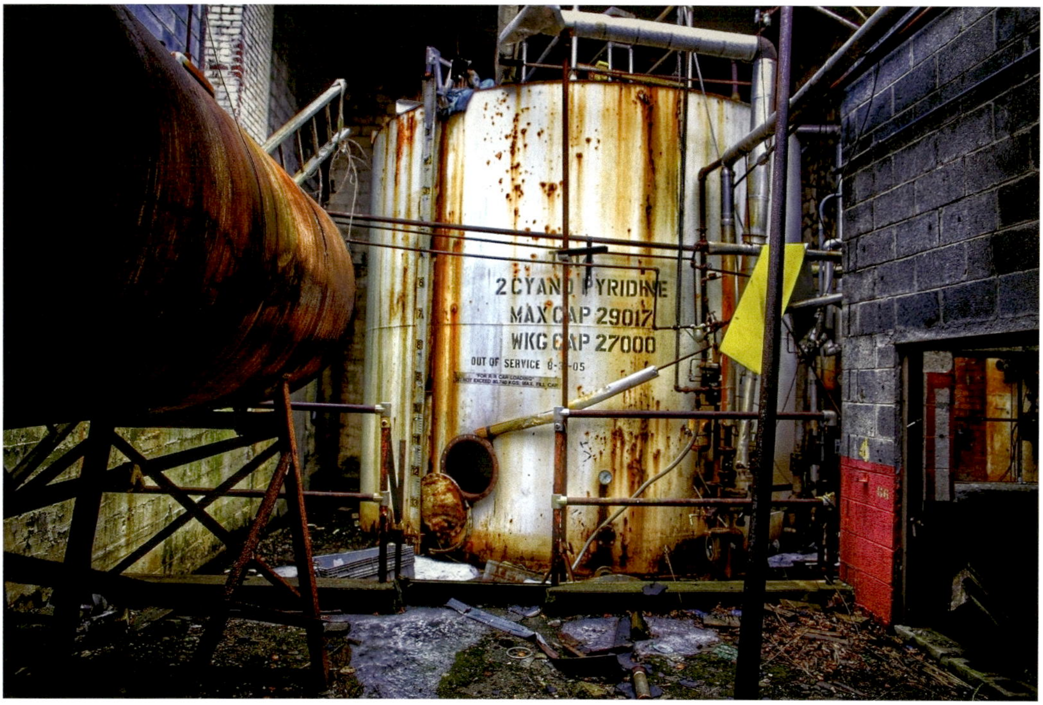

The Nepera Chemical Plant was the nation's second largest producer of pyridine, a chemical used in solvents, dandruff shampoos, and pesticides. (*Photo: Liz Cooke*)

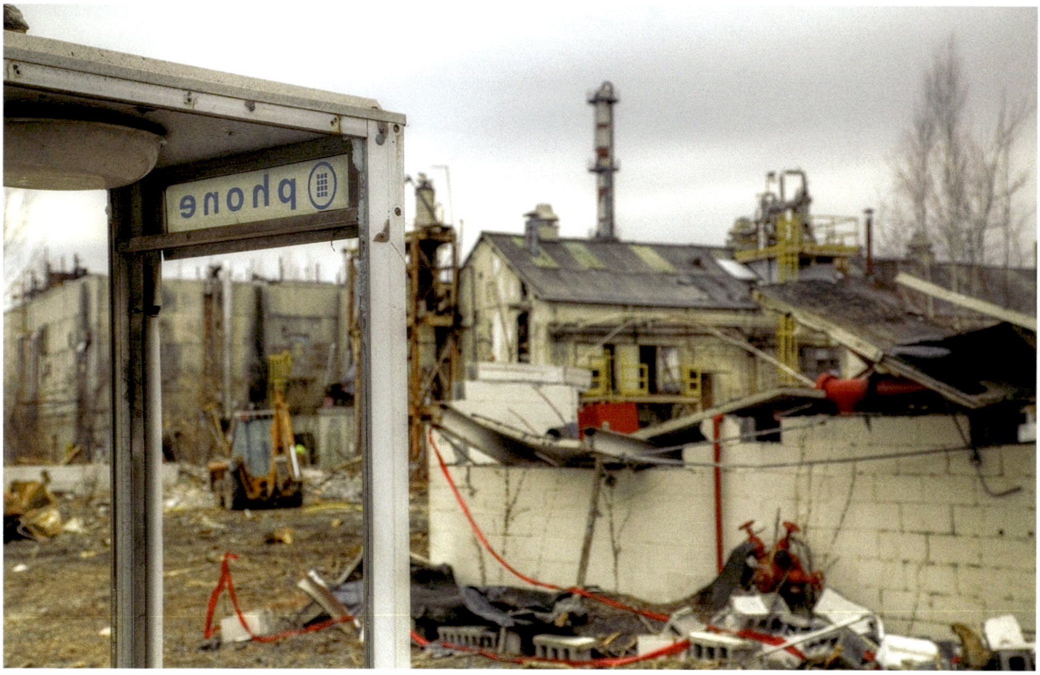

Before any reuse of the contaminated Nepera Chemical Plant can happen, an extensive review of environmental problems must be conducted. Dumping on the site included chemical solvents, pyridine, and mercury. Once employing 240 people, its fate remains uncertain. (*Photo: Andy Milford*)

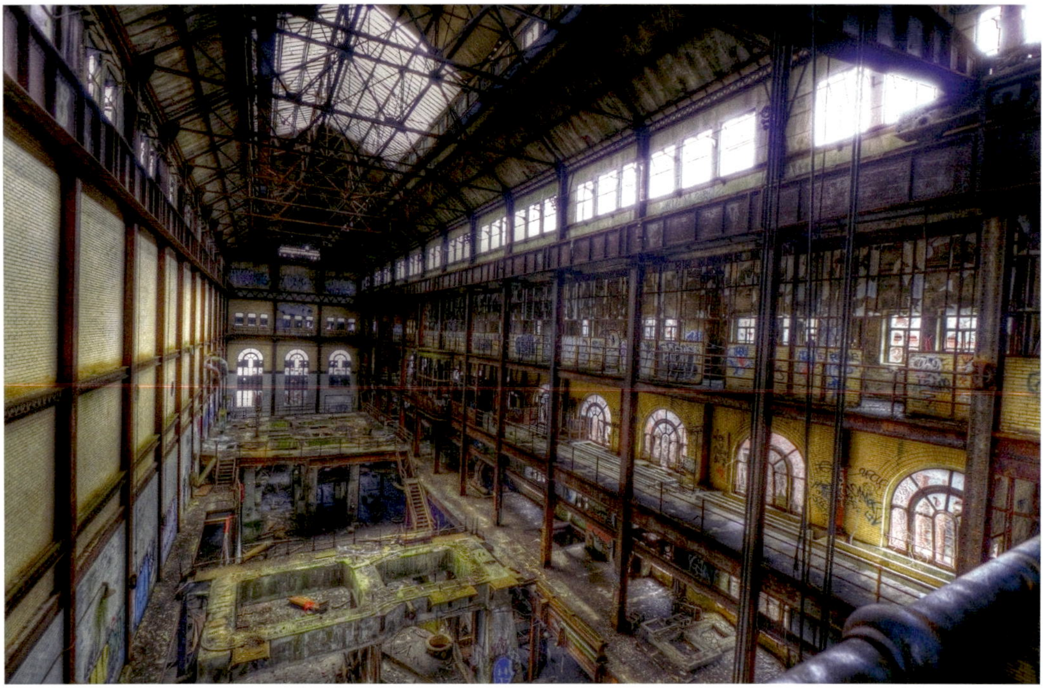

Inside the abandoned Yonkers Power Station in Yonkers, New York. Plans are under way for adaptive reuse as a headquarters for global climate solutions. (*Photo: Andy Milford*)

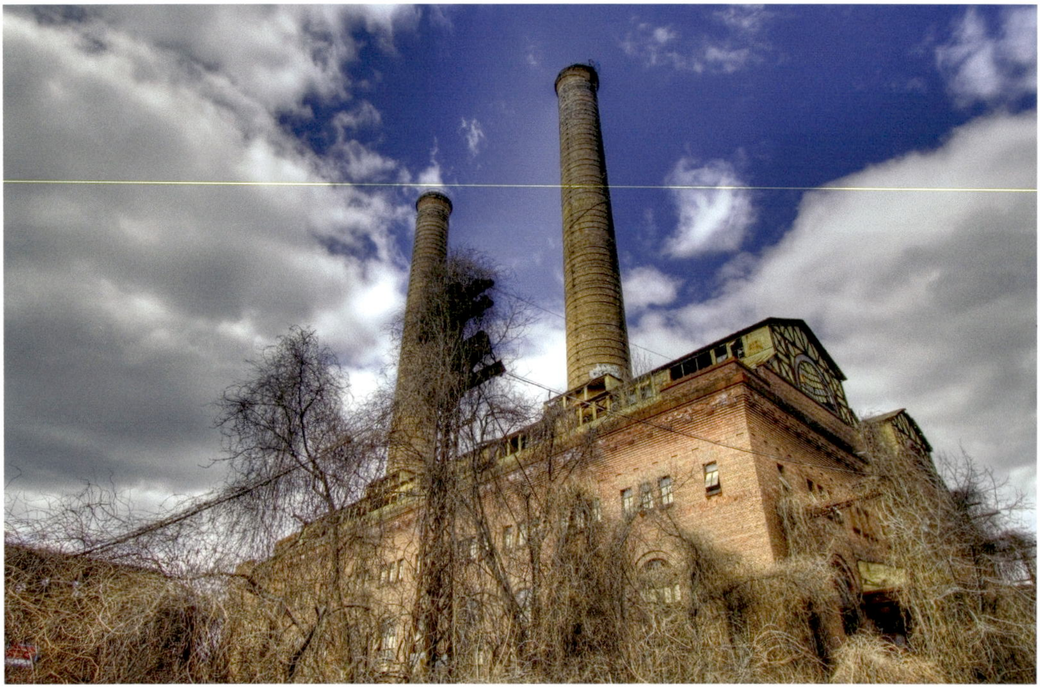

The Yonkers Power Station in Yonkers, New York, opened in 1907 and was used to meet the power needs of the New York Central & Hudson River Railroad. It stopped operation in 1963 and has been abandoned since. (*Photo: Andy Milford*)

York Central Railroad. The plant's Beaux-Arts design, characterized by its grand structure and intricate detailing, made it a local architectural landmark. However, as advancements in technology occurred, the plant was eventually decommissioned in the 1960s. After decades of abandonment and deterioration, efforts have been made to redevelop the site, aiming to transform the historic structure into a mixed-use complex focusing on global climate solutions.

The Lace Mill Factory (originally called the U.S. Lace Curtain Mill) was once a bustling hub of textile production in the Hudson Valley. Located in Kingston, New York, the factory specialized in producing high-quality lace products. As the mid-twentieth century approached, the shift towards cheaper overseas production led to a decline in local manufacturing and the Lace Mill, like many American factories, eventually closed its doors. Through community activism, the Lace Mill Factory has been repurposed into a vibrant artist community. Further into the Catskills, in Ellenville, New York, Imperial Schrade Knife was once known for its innovative products, including the switchblade. The factory was a major employer in Ellenville before closing its doors in 2004. In Wallkill, New York, the Borden Condensed Milk Factory has an optimistic future. Built in the mid-nineteenth century and wildly innovative for its time, the "condensary" was responsible for processing the milk of 5,000 cows from nearby farms. As demand for condensed milk "evaporated," the plant was no longer needed and was abandoned. It is now undergoing a major renovation, becoming an upscale hotel and event space. The former Williams Press publishing building in Menands, New York, had been abandoned for many years and is now a U-Haul facility.

Transportation played a vital role in the development of industry in the Hudson Valley and continues to do so. With its proximity to the Hudson River, ships of all sorts sailed from port to port. Railroads were a lifeline to New York City, Albany, and beyond. Later, cars became the dominant mode of transportation for individuals, with many people enjoying the beautiful roads of the Hudson Valley for trips in classic and vintage cars.

Throughout the region there are numerous abandoned railcars, automobiles, boats, ships, and even the hulking remains of the USS *Solace*, once a luxury passenger ship and later repurposed as "The Floating Hospital" during World War II. After its decommissioning, the ship was docked in Kingston, where it has languished for years.

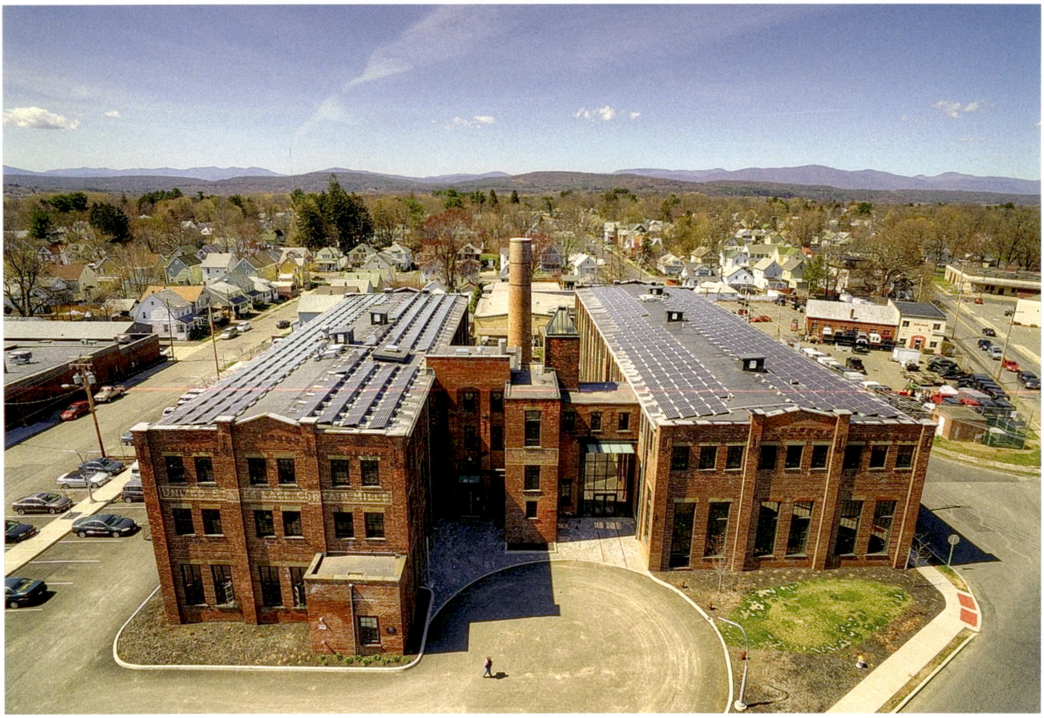

The Lace Mill Factory in Kingston, New York, was once a major producer of curtains and lace materials. After years of abandonment, it was transformed into an arts space and affordable housing. (*Photo: Andy Milford*)

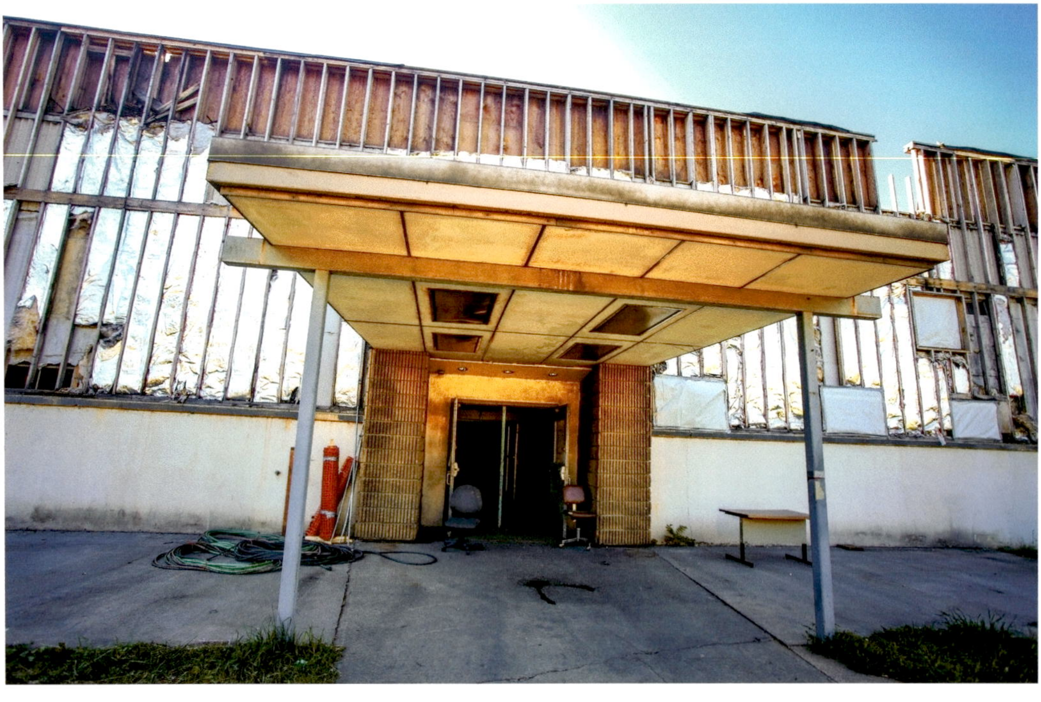

Imperial Schrade Knife in Ellenville, New York, was known for its innovative products including the switchblade and was a major area employer. It closed its doors in 2014. (*Photo: Liz Cooke*)

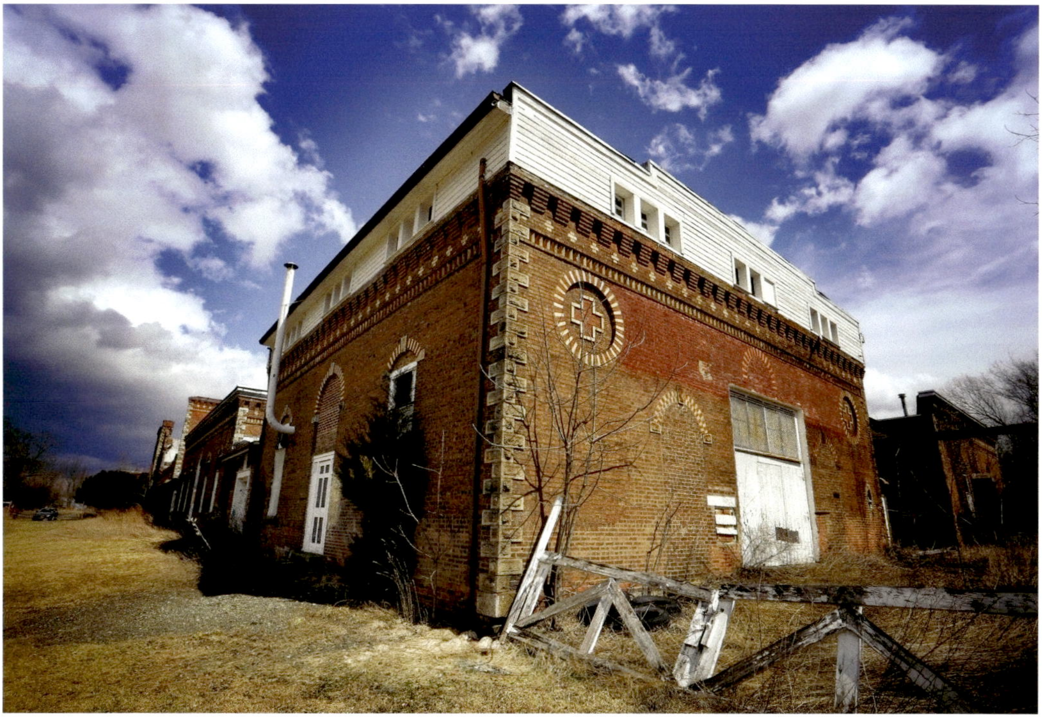

Corner view of the abandoned Borden Condensed Milk Factory (the "condensary") in Wallkill, New York. (*Photo: Liz Cooke*)

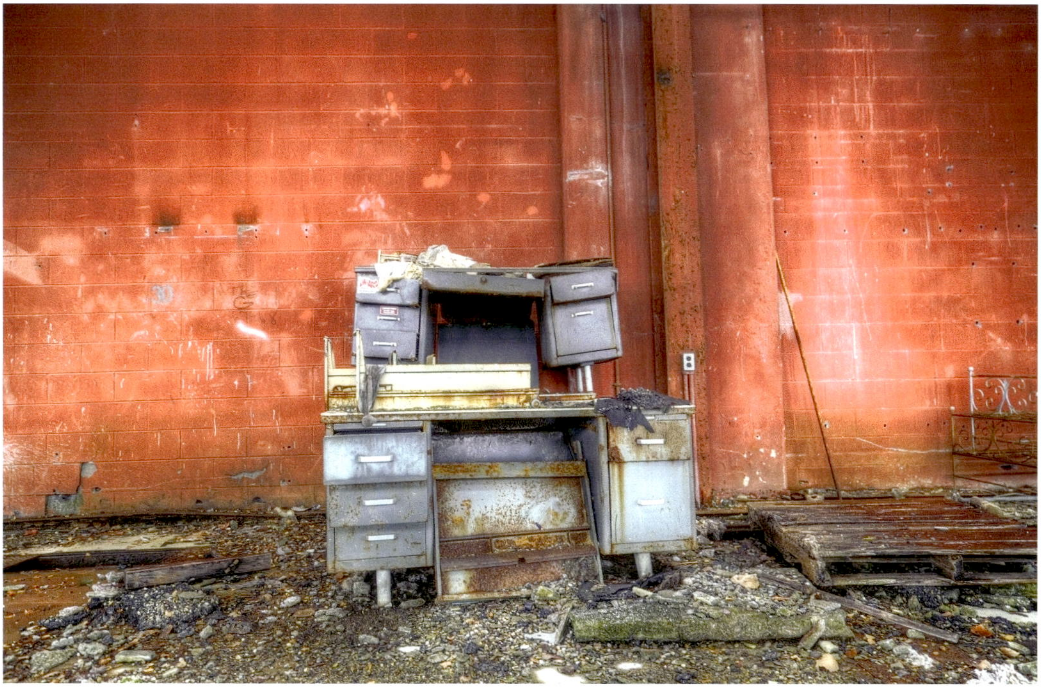

Two desks lean against the walls of the former Williams Press building in Menands, New York. (*Photo: Andy Milford*)

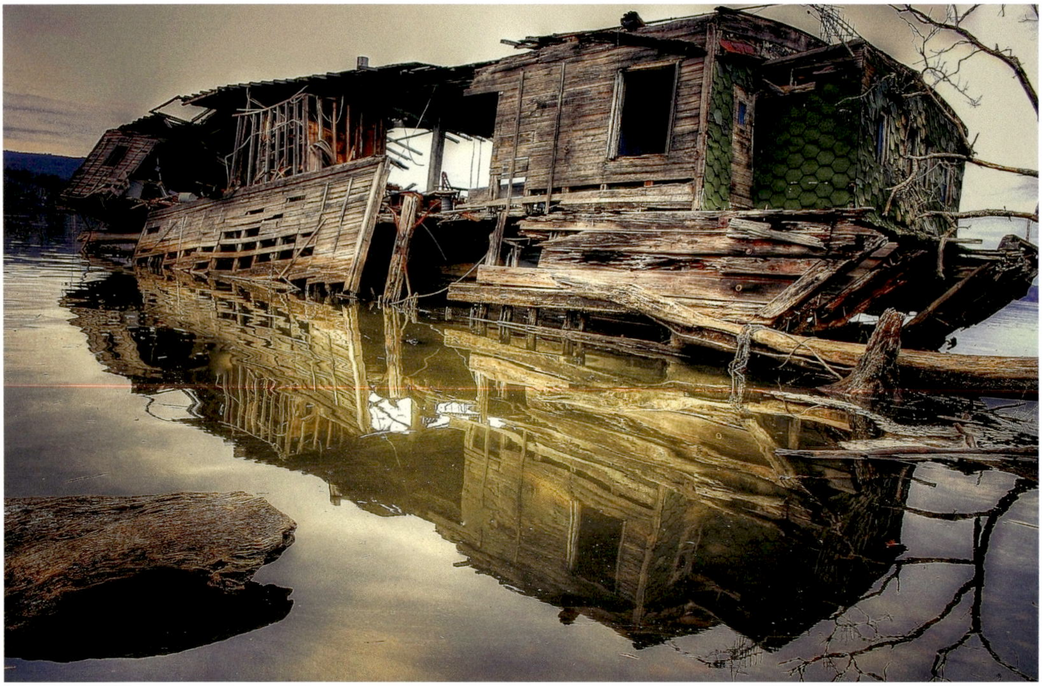

This decayed railroad barge (also called a "car float") sat in the Hudson River near Hyde Park, New York. After several years of hard winters and erosion, it completely vanished. (*Photo: Liz Cooke*)

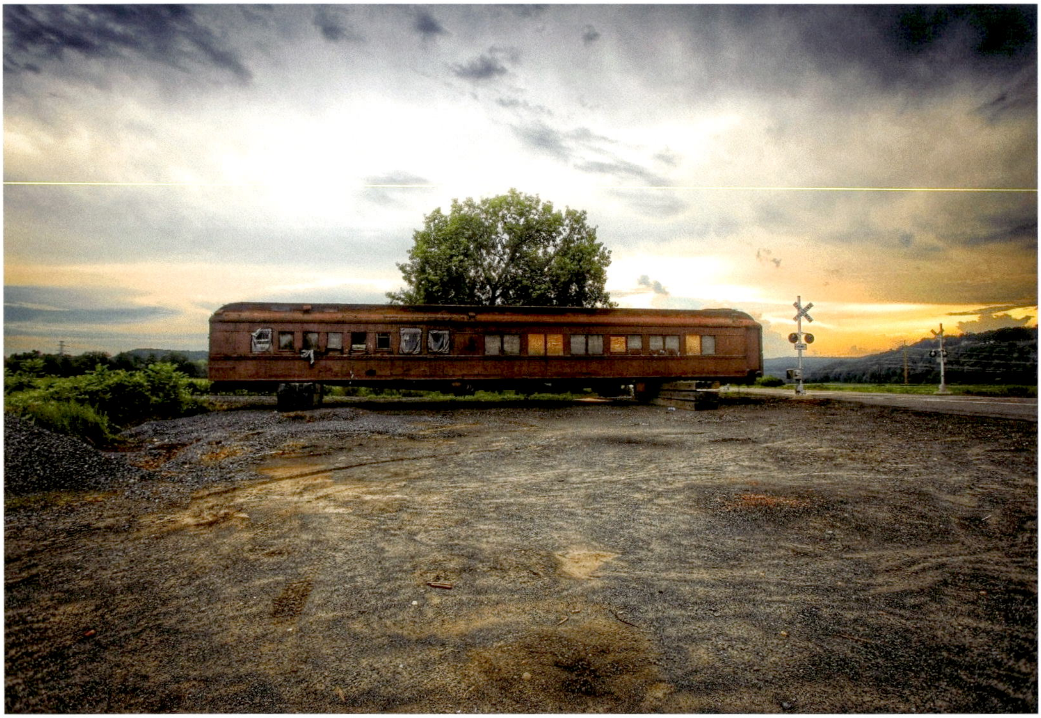

Sitting by the side of the road outside Kingston, New Yorkers, were several abandoned rail cars. The placement and purpose of these cars was never clear. After many years, they were moved. (*Photo: Liz Cooke*)

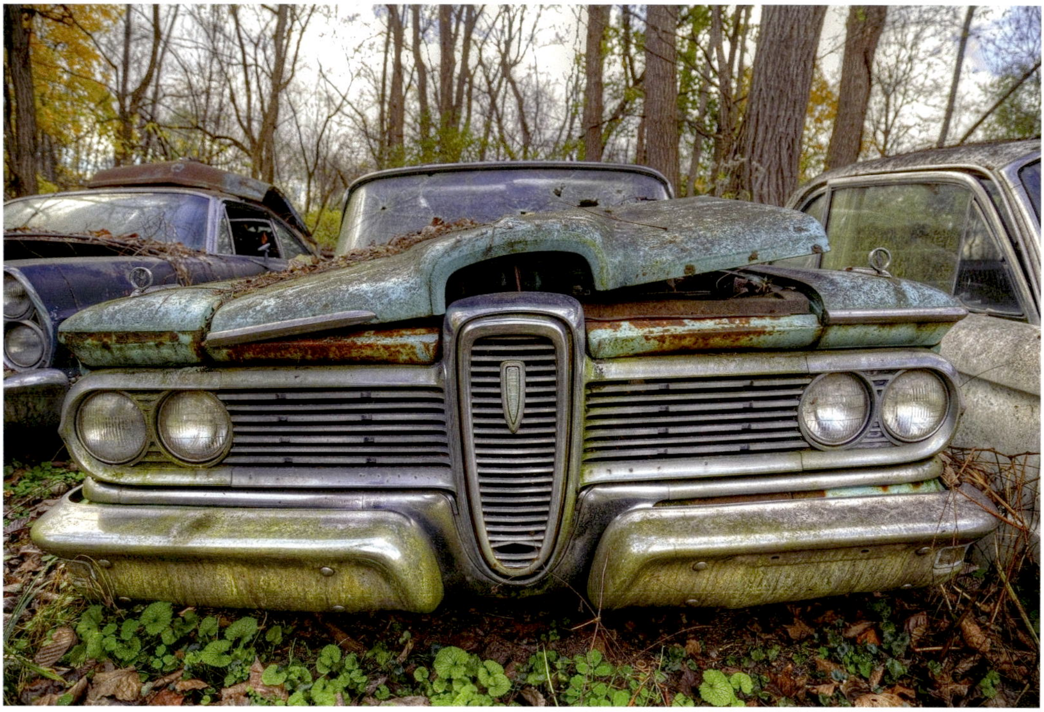

The woods in the Hudson Valley seem to hold more than their fair share of abandoned vehicles, but this vintage beauty is on the grounds of the Philmont, New York, Antique Auto Museum. (*Photo: Andy Milford*)

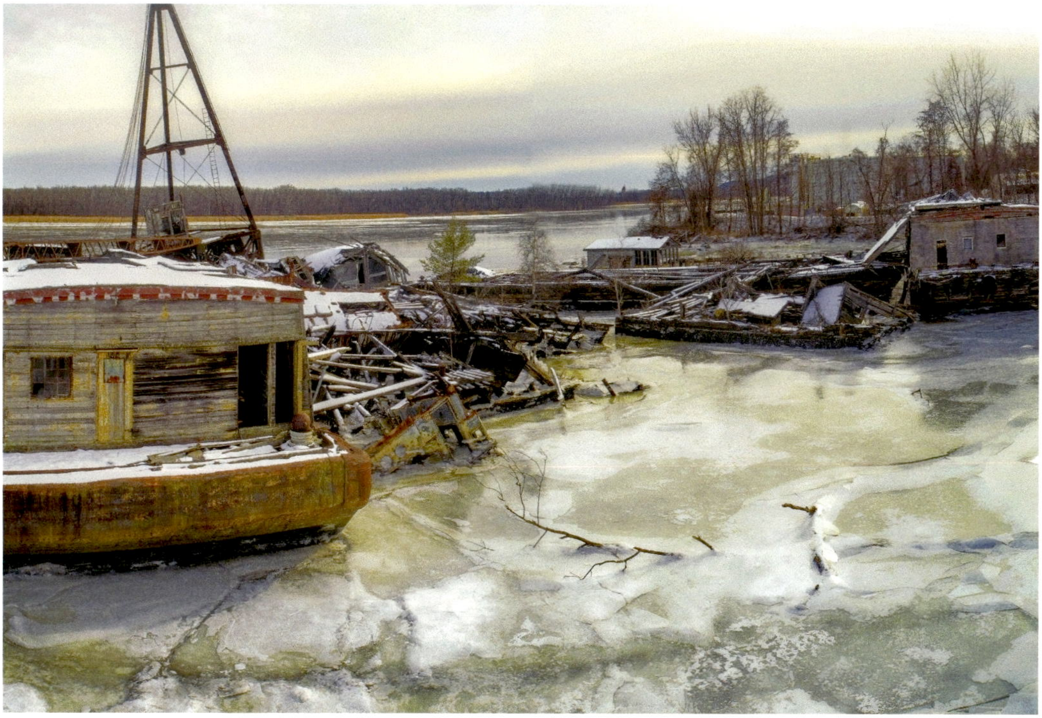

This railroad barge bearing the name "New York Central" is encased in ice in a "barge graveyard" outside Catskill, New York. (*Photo: Andy Milford*)

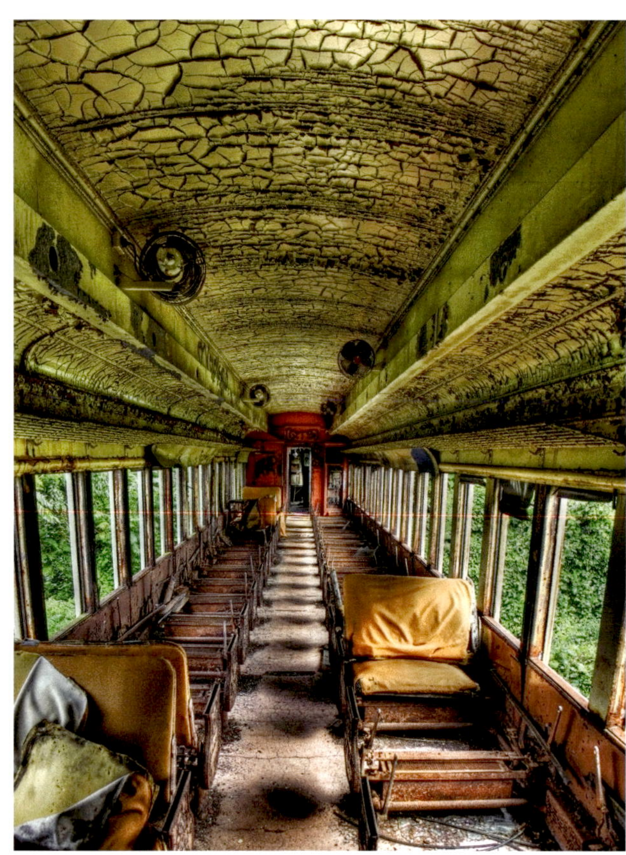

Left: Interior viewpoint of the "Kingston Express," an abandoned rail car outside Kingston, New York. (*Photo: Andy Milford*)

Below: Once known as the USS *Solace*, this former luxury passenger ship became "The Floating Hospital." It now sits in a Kingston, New York, shipyard. (*Photo: Andy Milford*)

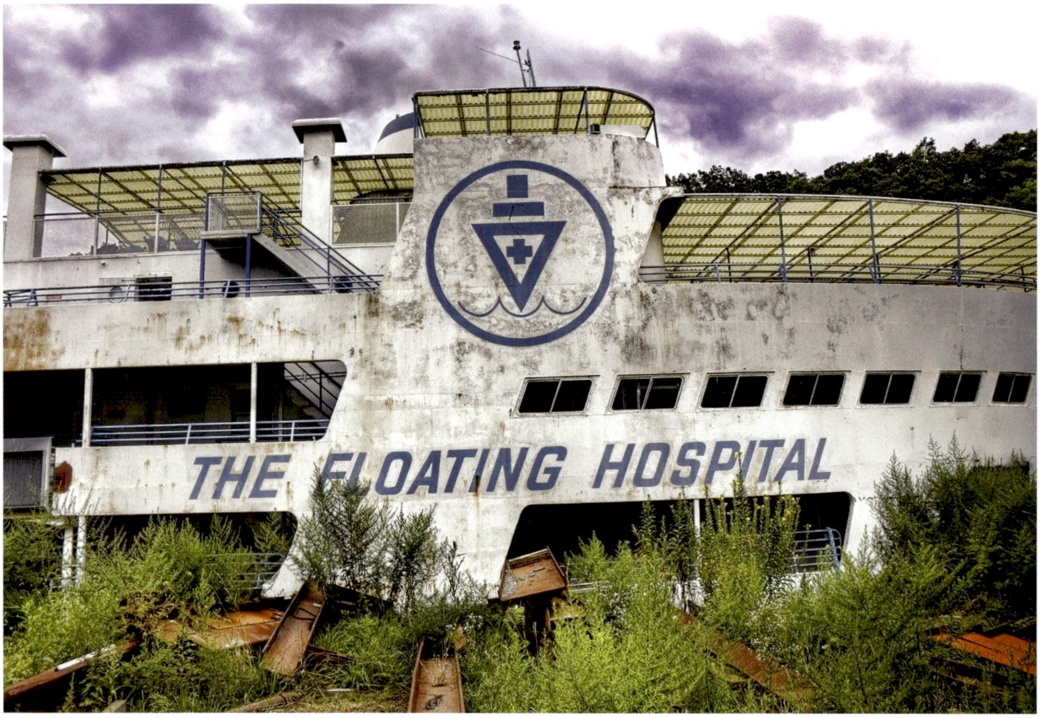

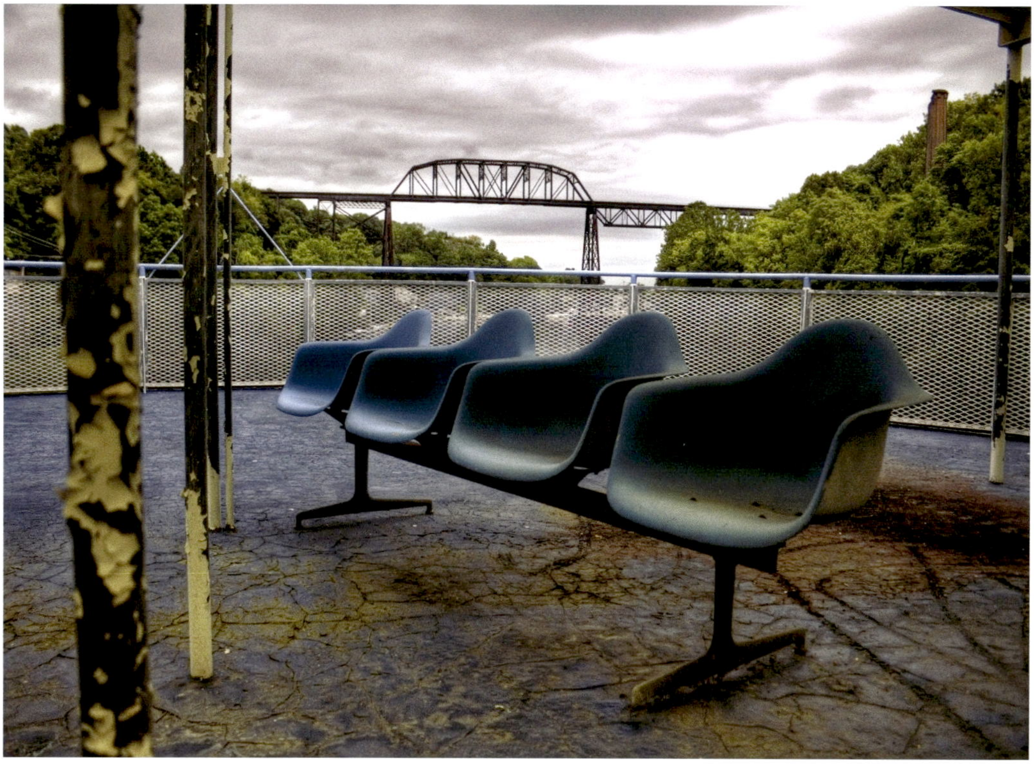

The deck level of "The Floating Hospital" in Kingston, New York, with the Wilbur Railroad Bridge in the background. (*Photo: Andy Milford*)

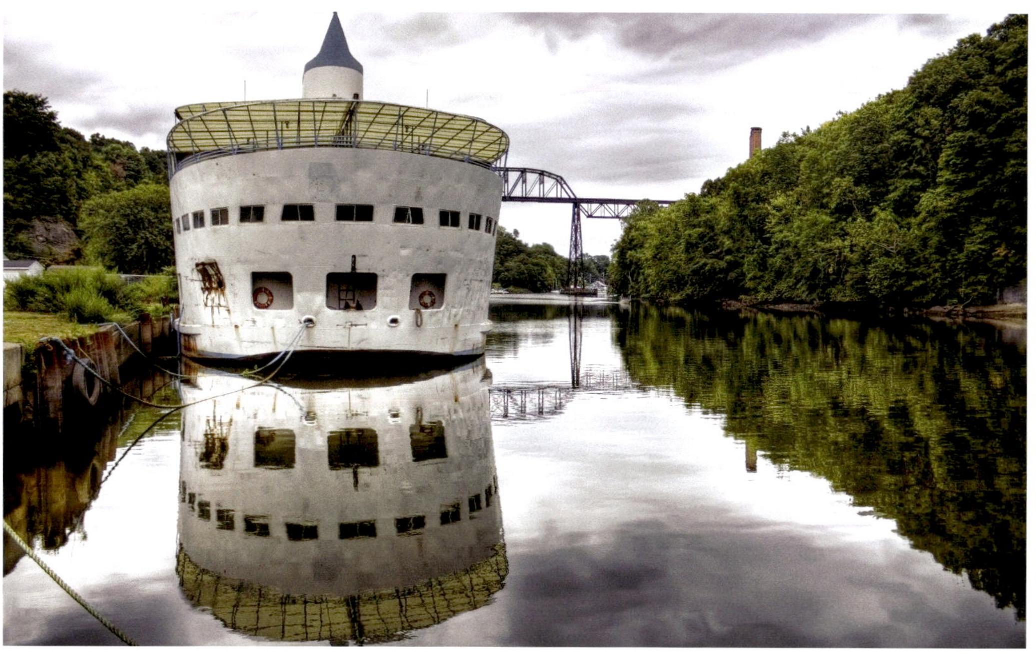

Located in the Rondout section of Kingston are the hulking remains of "The Floating Hospital," last used during World War II. (*Photo: Andy Milford*)

7

ROADSIDE ATTRACTIONS

With its agreeable climate and picturesque scenery, the Hudson Valley has long been a retreat for city dwellers seeking respite from city life and local families looking for easy escapes from everyday stresses. From funky diners and drive-in theaters to an ersatz Wild West village and a rustic retreat for New York City sanitation workers, the region provided a unique blend of entertainment, education, and escapism that kept people coming back for more. With so many offerings, it must have seemed like it would last forever, but as the decades passed, affordable air travel and more luxurious destinations inspired visitors to seek their fun elsewhere, leaving many attractions in a state of decay and abandonment.

The Catskill Game Farm, established in 1933, quickly became a beloved destination for families. Located in Catskill, New York, it was the first privately owned zoo in the United States. By the 1950s, it had grown into a sprawling 914-acre park, home to over 2,000 animals. During its peak in the mid-twentieth century, the Game Farm symbolized an era of post-war prosperity and optimism. The burgeoning middle class had more disposable income and leisure time, allowing for frequent family outings. However, by the 1980s, the appeal of the Game Farm began to wane. Increasing competition from modern, interactive theme parks, coupled with rising operational costs and evolving public attitudes towards animal captivity, led to its closure in 2006. The property was purchased in the 2000s and is now operating as a unique bed and breakfast.

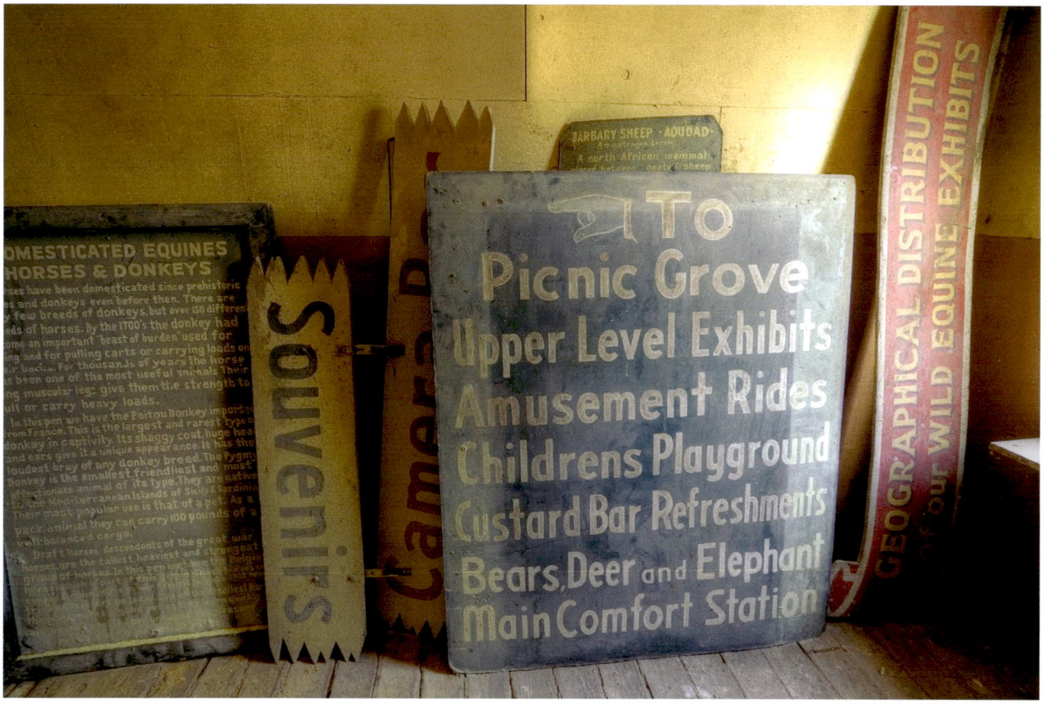

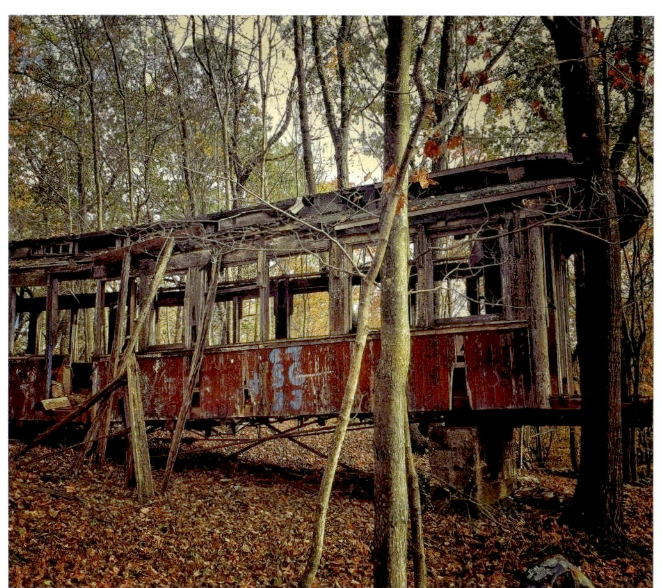
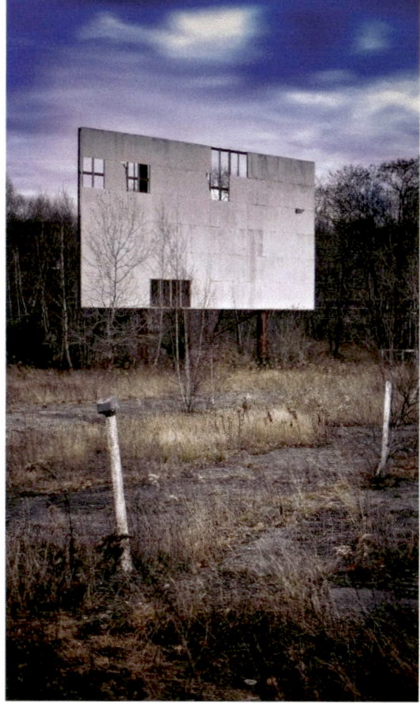

Signs invite visitors to explore the grounds of the Catskill Game Farm, once a popular family destination in Catskill, New York. (*Photo: Andy Milford*)

Above left: This old trolley car was once a rustic accommodation for vacationing New York City Sanitation workers in an area known as "Sanitas Hills" in Holmes, New York. Once a family-friendly destination with accommodations and all manner of recreation, it has been abandoned for many years. (*Photo: Liz Cooke*)

Above right: The abandoned Mountain Drive-In, located in Liberty, New York, offered three screens for visitors' viewing pleasure. It opened in 1949 and closed in 1997. (*Photo: Liz Cooke*)

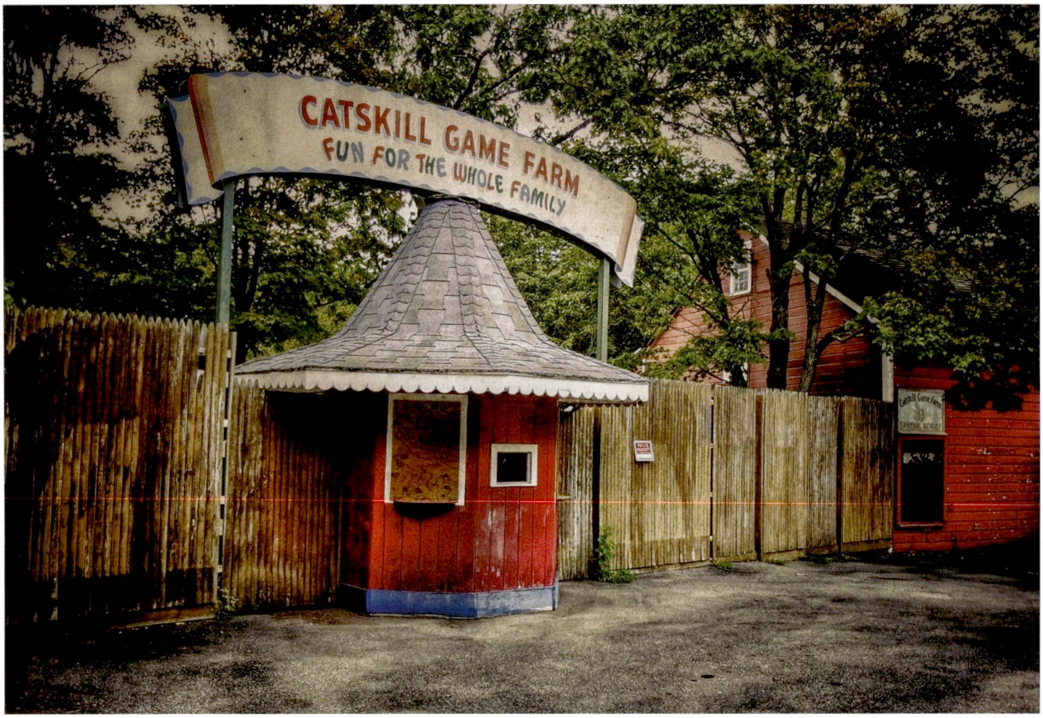

The entrance to the Catskill Game Farm promised "fun for the whole family." Opened in 1933, the Game farm attracted thousands of visitors each year before closing in 2006. The site is now a boutique inn. (*Photo: Liz Cooke*)

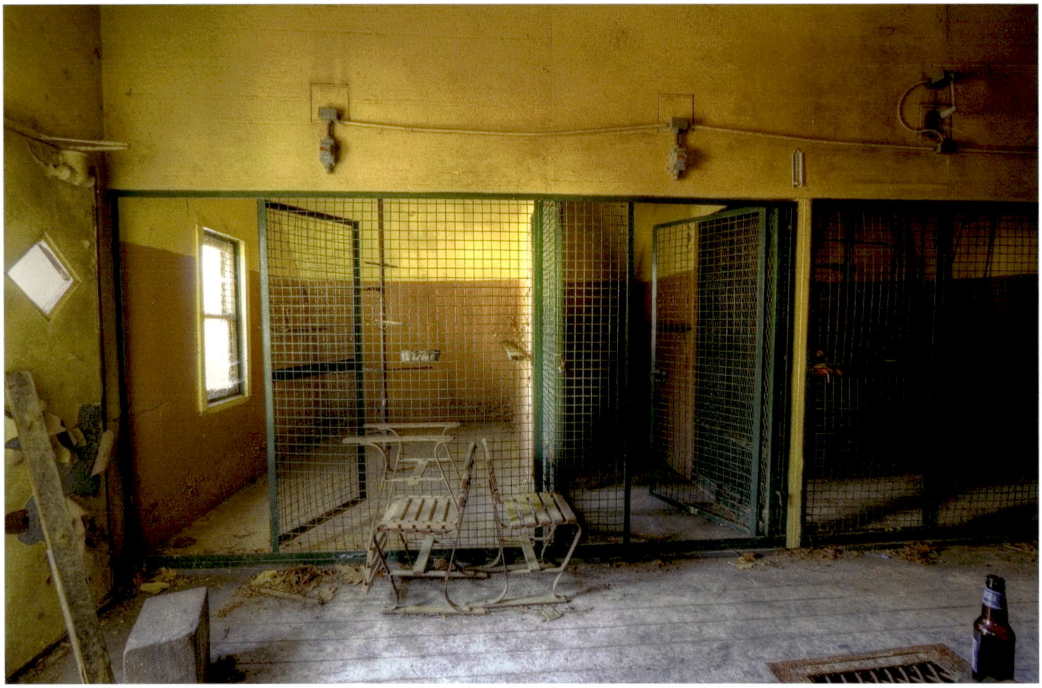

Animal enclosures at the Catskill Game Farm allowed visitors to mingle with exotic cats, giraffes and many other animals. At its height, over 2,000 animals called the Catskill Game Farm home. (*Photo: Andy Milford*)

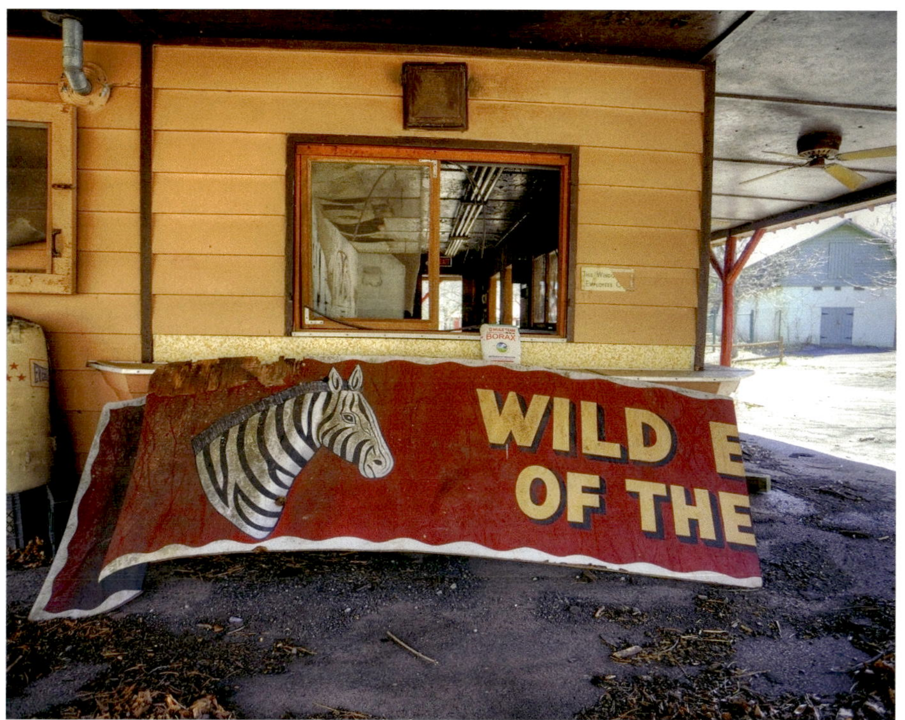

Signage for the "Wild Equines of the World," lies on the ground of the Catskill Game Farm. (*Photo: Andy Milford*)

Carson City and Indian Village offered a different kind of escapism. Opening in 1958 near Catskill, New York, this theme park recreated the adventurous spirit of the American West. Visitors could immerse themselves in staged gunfights, Native American dances, and Wild West shows, all set against a backdrop of meticulously constructed Western-style buildings. The popularity of Carson City peaked in the 1960s and 1970s. However, the park faced several challenges as cultural sensibilities shifted. The portrayal of Native Americans and the Wild West became subjects of scrutiny and criticism. These factors, combined with economic difficulties, led to the park's closure in 1997.

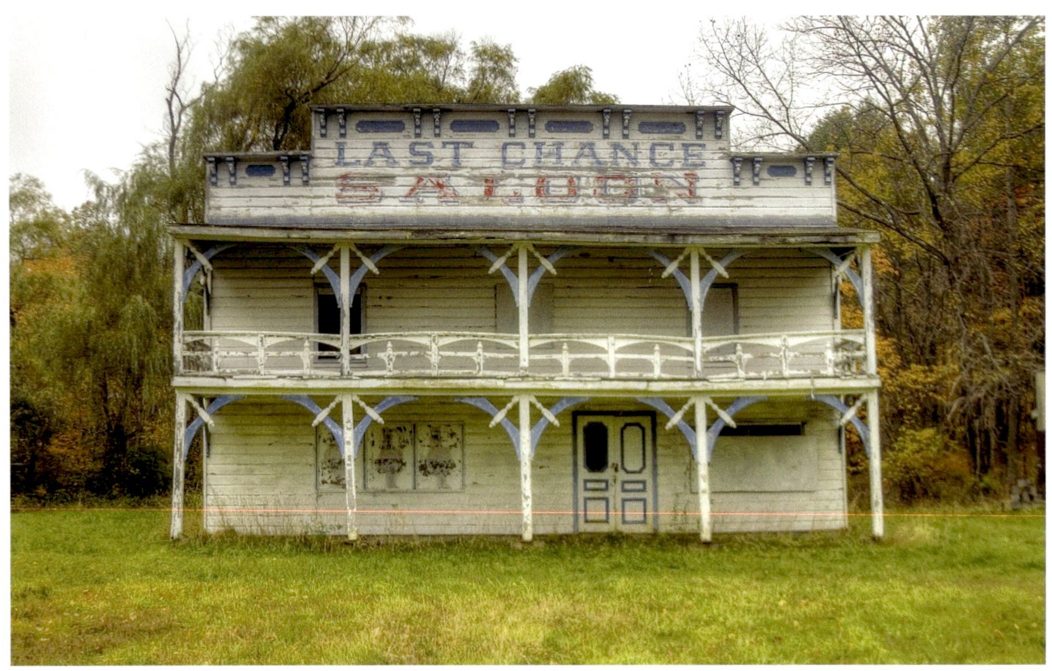

The Last Chance Saloon was a popular attraction at Carson City and Indian Village, a wild-west themed amusement park in Catskill, New York. Crowds cheered at staged gunfights right outside the saloon. (*Photo: Andy Milford*)

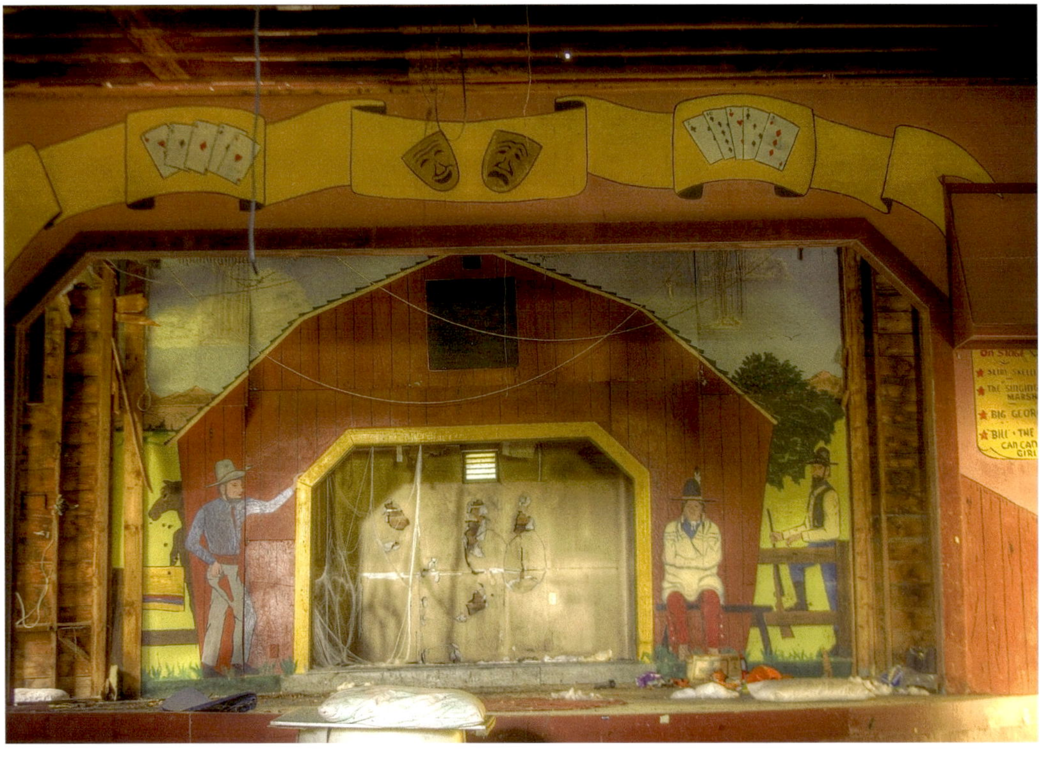

The portrayal of native peoples in the American West may have led to Carson City and Indian Village's closure in 1997. (*Photo: Andy Milford*)

8

RUINS

Exploring the ruins of the Hudson Valley is a chance to uncover the stories of the past hidden among crumbling walls and overgrown paths. As you wander through the rugged terrain, you cannot help but be struck by the beauty of these forgotten places. Fascinating ruins are scattered throughout the landscape, hidden in the forests, nestled in the valleys, and perched on the hillsides, waiting to be discovered. Some ruins are well-preserved, their walls still standing, while others are mere shells of their former selves, their roofs collapsed and their windows nothing but empty holes.

Among the region's most captivating ruins are the remains of the Overlook Mountain House. Built and rebuilt three times beginning in 1871, the Overlook Mountain House, like many other Catskill retreats, was built to allow wealthy city dwellers a chance to experience nature in a luxurious setting. The first hotel was destroyed by fire in 1875. The second hotel lost business to competing mountain hotels and resorts—many at lower, and therefore more accessible, altitudes. In 1923, this incarnation was also destroyed by fire. The third incarnation was built of concrete but was never completed. The hotel was abandoned in the 1940s and destroyed by fire in the 1960s. What remains is little more than a hollow shell.

Another intriguing ruin is Bannerman Castle, located on Pollepel Island in the Hudson River. Built by Francis Bannerman VI beginning in 1901, the castle served as a storage facility for Bannerman's vast collection of military surplus equipment and munitions. Bannerman, a Scottish immigrant and successful arms dealer, designed the castle to resemble the fortified structures of his homeland. The island, bought in 1900, was transformed into a private arsenal, with the distinctive castle becoming an iconic landmark. However, after Bannerman's death in 1918 and

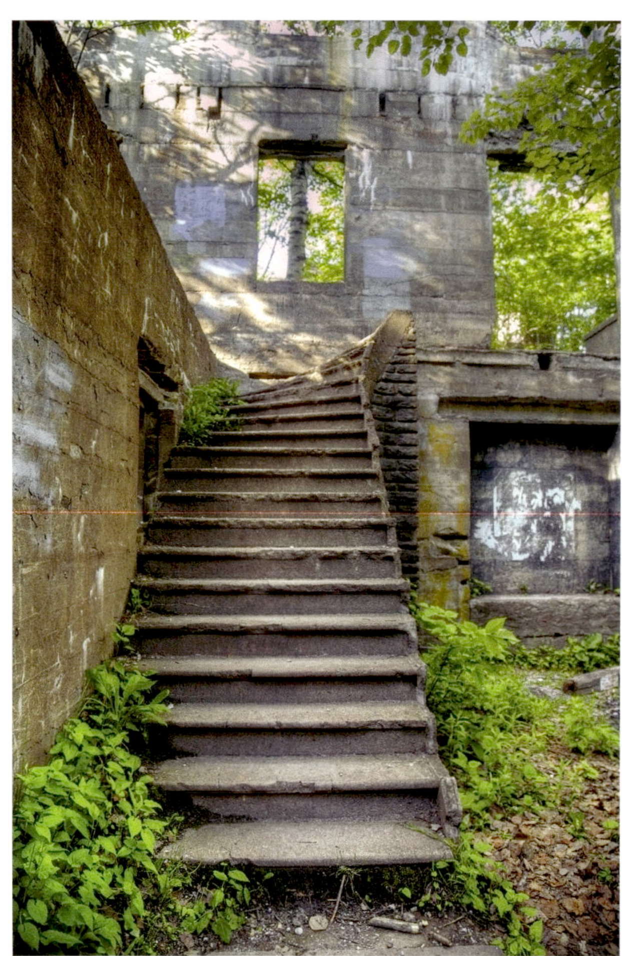

Left: The ruins of Overlook Mountain House near the summit of Overlook Mountain outside Woodstock, New York, are all that remains of the popular nineteenth-century mountain retreat. (*Photo: Andy Milford*)

Below: The skeletal remains of the Overlook Mountain House as seen from above. (*Photo: Andy Milford*)

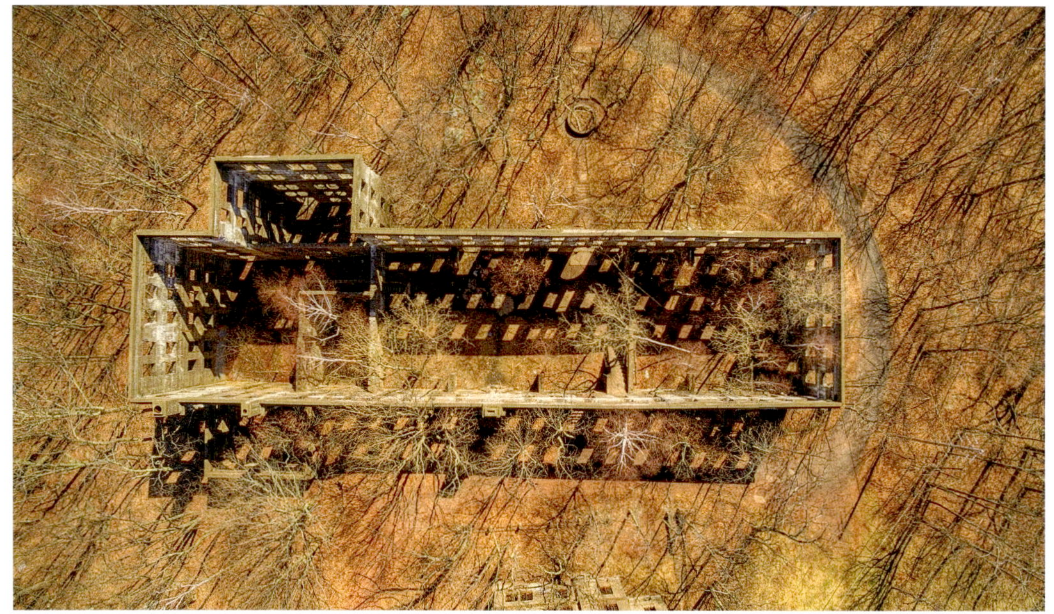

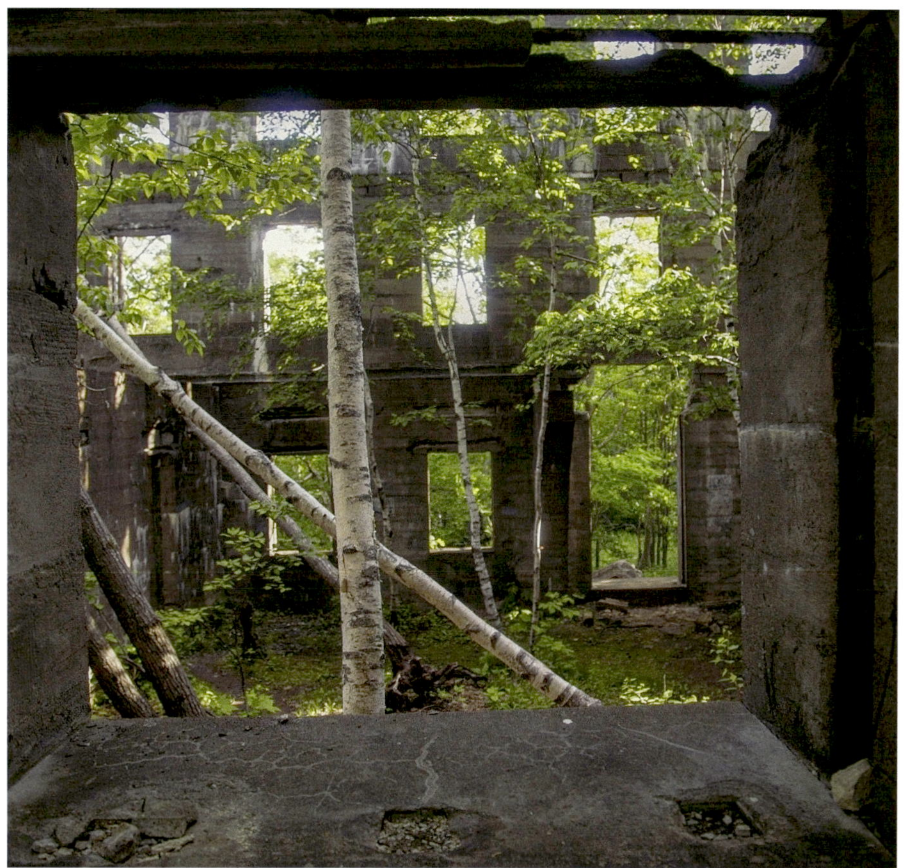

Many hikers brave the trail to reach the ruins of the Overlook Mountain House. While very little remains of the main house, windows give a sense of the views visitors would have enjoyed. (*Photo: Andy Milford*)

a catastrophic explosion in 1920, the castle's use declined. Over the decades, the structure fell into disrepair, exacerbated by a major fire in 1969. Today, Bannerman Castle stands as a romantic ruin, accessible via guided tours, and is preserved by the Bannerman Castle Trust as a historical and cultural site.

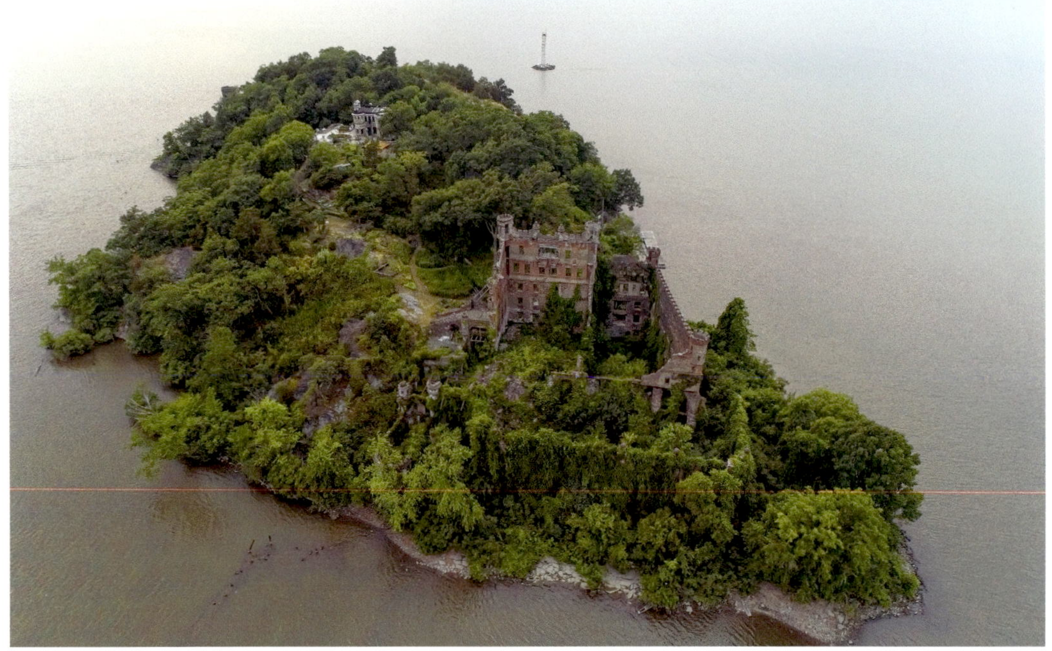

One of the most iconic ruins in the Hudson Valley is Bannerman Castle on Pollepel Island. The island was purchased by Francis Bannerman VI in 1900 as a place to store his massive cache of surplus munitions. Buildings designed by Bannerman were used to store munitions and as a private residence. (*Photo: Andy Milford*)

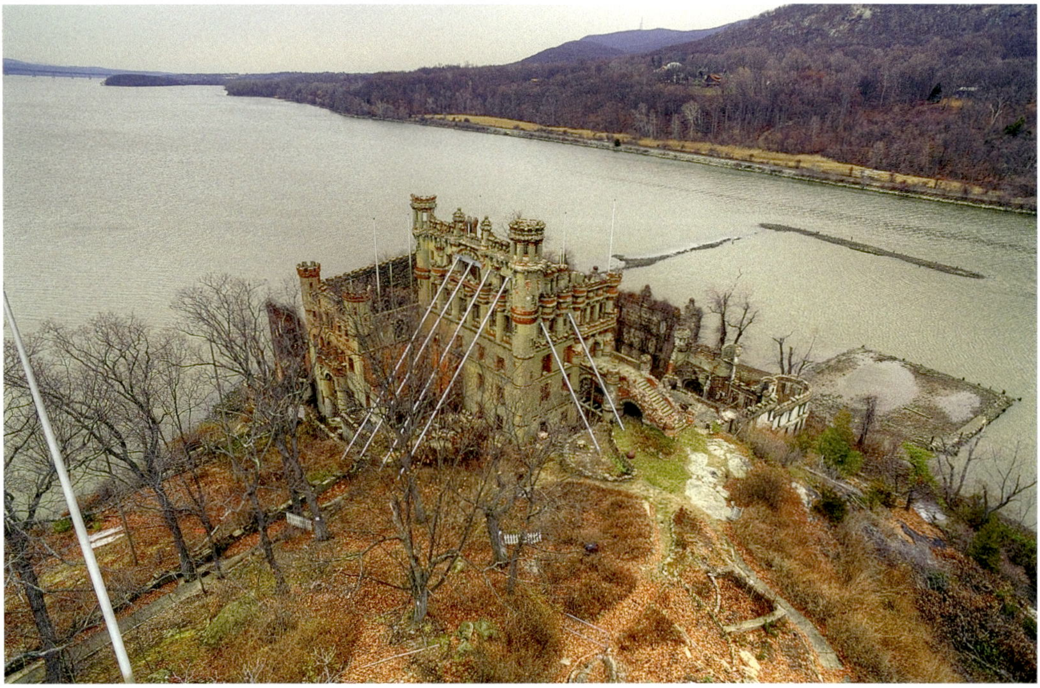

The ruins of Bannerman Castle on Pollepel Island in the Hudson River as seen from above. Little remains of the original structure. The property is now owned by the New York State Office of Parks, Recreation and Historic Preservation. (*Photo: Andy Milford*)

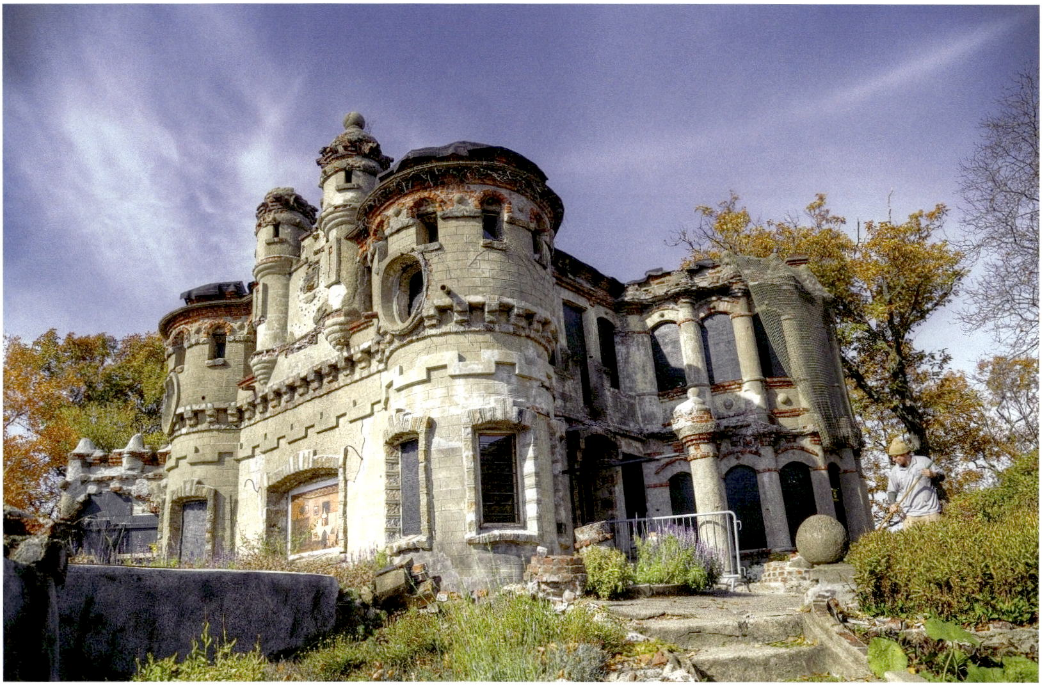

This smaller "castle" was designed by Francis Bannerman VI as his private residence. It is largely intact while other structures on the island have been destroyed by fire and the forces of nature. (*Photo: Andy Milford*)

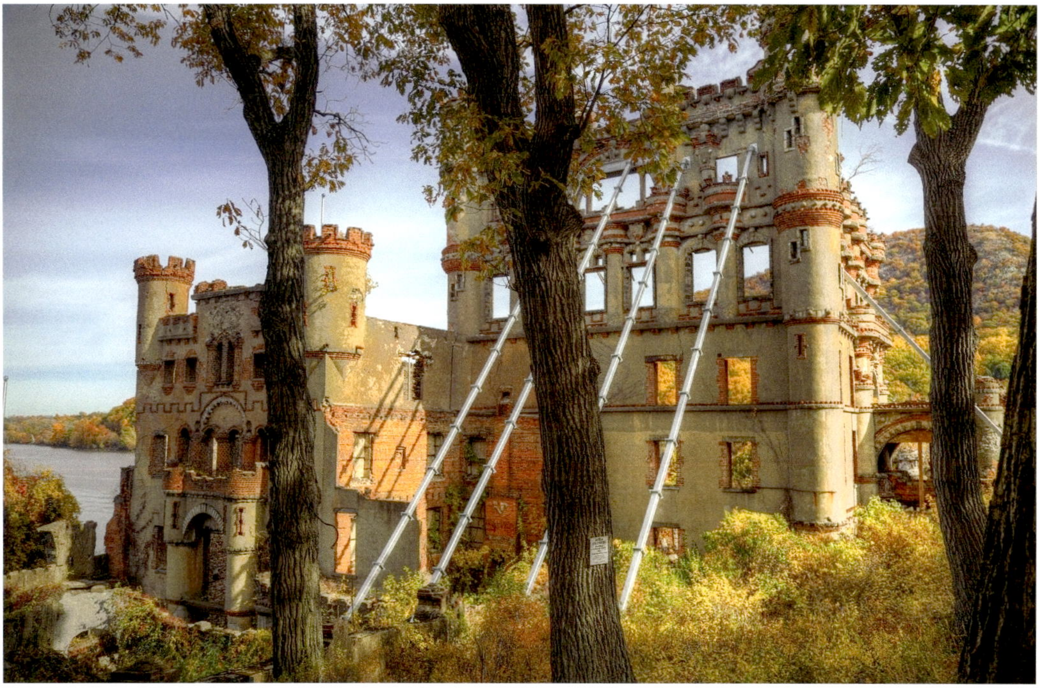

Despite its fragile state, Bannerman Castle still has the power to capture the imagination. Tours are conducted and visitors are invited to attend lectures and exhibitions during the warmer months. (*Photo: Andy Milford*)

EPILOGUE

We hope you have enjoyed traveling with us on this journey through New York's forsaken past. And we hope we have inspired you to look for the spaces and places others miss. We would be remiss if we did not end with a few cautions. Many of the places described in this book are off-limits and no attempt should be made to enter without permission. Many places have decayed at alarming rates and present real hazards. We encourage photographers and aspiring "abandonistas" to exercise good judgment and in most cases, shoot exteriors only. To protect these fragile places, we did not give specific locations. Some sites are well known, and we made no attempt to conceal their location. If we had one closing remark, it would be this: please take good care as you explore the forsaken places in your own hometown. And please take good care of the places you find.

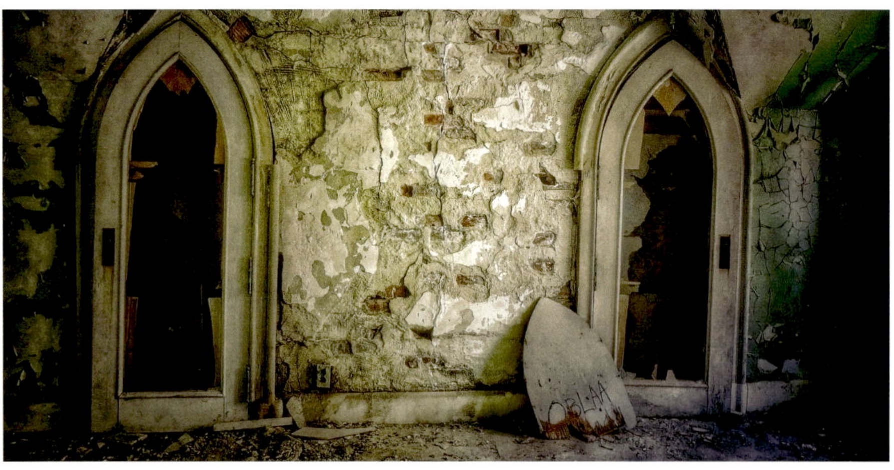

Architectural detail from Dundas Castle, Roscoe, New York. (*Photo: Liz Cooke*)